Peat, F. David, 1938-

From certainty to uncertainty : the story of science and ideas in the twentieth century

FROM CERTAINTY TO UNCERTAINTY

Also by F. David Peat

Blackfoot Physics: A Journey into the Native American Universe

Seven Life Lessons of Chaos: Timeless Wisdom from the Science of Change (with John Briggs)

The Blackwinged Night: Creativity in Nature and Mind

Science, Order, and Creativity (with David Bohm)

Infinite Potential: The Life and Times of David Bohm

In Search of Nikola Tesla

Who's Afraid of Schrödinger's Cat? An A-to-Z Guide to All the New Science Ideas You Need to Keep Up with the New Thinking (with Ian Marshall and Danah Zohar)

Glimpsing Reality: Ideas in Physics and the Link to Biology (edited, with Paul Buckley)

The Philosopher's Stone: Chaos, Synchronicity, and the Hidden Order of the World

Quantum Implications: Essays in Honour of David Bohm (edited, with Basil Hiley)

Einstein's Moon: Bell's Theorem and the Curious Quest for Quantum Reality

Superstrings and the Search for the Theory of Everything

Turbulent Mirror: An Illustrated Guide to Chaos Theory and the Science of Wholeness (with John Briggs)

Cold Fusion: The Making of a Scientific Controversy

Artificial Intelligence: How Machines Think

Synchronicity: The Bridge Between Matter and Mind

Looking Glass Universe: The Emerging Science of Wholeness (with John Briggs)

The Armchair Guide to Murder and Detection

The Nuclear Book

FROM CERTAINTY TO UNCERTAINTY

The Story of Science and Ideas in the Twentieth Century

F. DAVID PEAT

JOSEPH HENRY PRESS
WASHINGTON, D.C.

Joseph Henry Press • 2101 Constitution Avenue, N.W. • Washington, D.C. 20418

The Joseph Henry Press, an imprint of the National Academy Press, was created with the goal of making books on science, technology, and health more widely available to professionals and the public. Joseph Henry was one of the founders of the National Academy of Sciences and a leader in early American science.

Library of Congress Cataloging-in-Publication Data

Peat, F. David, 1938-
From certainty to uncertainty : the story of science and ideas in the twentieth century / F. David Peat.
p. cm.
Includes index.
ISBN 0-309-07641-2 (hard)
1. Physics—Philosophy. 2. Certainty. 3. Chaotic behavior in systems. 4. Physics—History—20th century. I. Title: Story of science and ideas in the twentieth century. II. Title.
QC6 .P33 2002
530'.09'04—dc21

2002001482

Cover art: Diego Rodriguez Velazquez, Las Meninas (detail), copyright Erich Lessing/Art Resource, NY (left side); Michele de la Menardiere, Homage to Las Meninas (right side).

Printed in the United States of America.

For Alessandro

CONTENTS

PREFACE

Que sais-je? (What do I know?) *Montaigne*

The first year of a new century always appears auspicious. The year 1900 was no exception. Americans welcomed it in with the three Ps: Peace, Prosperity, and Progress. It was the culmination of many outstanding achievements and looked forward, with great confidence, to a century of continued progress. The twentieth century would be an age of knowledge and certainty. Ironically it ended in uncertainty, ambiguity, and doubt. This book is the story of that change and of a major transformation in human thinking. It also argues that, while our new millennium may no longer offer certainty, it does hold a new potential for growth, change, discovery, and creativity in all walks of life.

On April 27, 1900, Lord Kelvin, the eminent physicist and president of Britain's Royal Society, addressed the Royal Institution, pointing out "the beauty and clearness of the dynamical theory." Finally Newton's physics had been extended to embrace all of physics, including both heat and light. In essence, everything that could be known was, in principle at least, already known. The president could look ahead to a new century with total conviction. Newton's theory of motion had been confirmed by generations of scientists, and it explained everything from the orbits of the planets to the times of the tides, the fall of an apple, and the path of a projectile. What's more, during the preceding decades James Clerk Maxwell had established a definitive theory of light. Taken together, Newton's and Maxwell's two theories appeared to be capable of explaining every phenomenon in the entire physical universe.

Yet the cusp of the twentieth century presents us with an irony. 1900 was a year of great stability and confidence. It saw the consolidation and summing up of many triumphs in science, technology, engineering, economics, and diplomacy. As Senator Chauncey Depew of New York put it, "There is not a man here who does not feel 400 percent bigger in 1900 than he did in 1896, bigger intellectually, bigger hopefully, bigger patriotically," while the Reverend Newell Dwight Hillis claimed, "Laws are becoming more just, rules more humane; music is becoming sweeter and books wiser." Yet, at that very moment other thinkers, inventors, scientists, artists, and dreamers, including Max Planck, Henri Poincaré, Thomas Edison, Guglielmo Marconi, Nikola Tesla, the Wright brothers, Bertrand Russell, Paul Cézanne, Pablo Picasso, Marcel Proust, Sigmund Freud, Henry Ford, and Herman Hollerith were conceiving of ideas and inventions that were to transform the entire globe.

1900 was the year in which flash photography was invented and speech was first transmitted by radio. Arthur Evans discovered evidence of a Minoan culture and the United States backed its paper currency with gold. Once the Gold Standard had been adopted, was there anything that could stand in the way of a greater degree of confidence in the future of their world?

1900 also represents the culmination of a period of rapid discovery. In the two previous years the Curies had discovered radium and J. J. Thomson the electron. Von Linde had liquefied air and Aspirin had been invented. Edison's Vitascope together with the magnetic recording of sound heralded the age of the movies.

Thanks to Nikola Tesla's inventions in alternating current, the city of Buffalo was receiving electrical power generated by Niagara Falls. Count von Zeppelin constructed an airship, the Paris Metro opened, and London saw its first motorbus. By 1902, the transmission of data by telephone and telegraph was already well established, and the first faxed photographs were being transmitted.

1900 also saw a link between Britain's Trades Union Congress and the Independent Labour Party, a move that would eventually lead to the establishment of the welfare state. With such a dream of social improvement people seemed justified in believing that the future would provide better housing, education, and health services. Homelessness would be a thing of the past and, while those thrown out of work would need to tighten their belts a little, they would be supported by the welfare state and would no longer face suffering and hardship.

Europe also experienced a great sense of stability in 1900. Queen Victoria, who had ruled since 1837, was still on the throne. She had become known as "the Grandmother of Europe," since her grandchildren were now part of the European monarchy. Indeed all of the European kings and queens, as well as the Russian royal family, were a part of a single international family presided over by Victoria. It was for this reason, diplomats believed, there would never be a war within Europe.

On May 18, 1899, at the prompting of Czar Nicholas II's minister of foreign affairs, 26 nations met at The Hague for the world's first peace conference. There they established an International Court to arbitrate in disputes between nations. The conference outlawed poison gases, dumdum bullets, and the discharge of bombs from balloons. Wars and international conflicts would be things of the past. The world itself was moving toward a new golden age in which science and technology would be put to the service of humanity and world peace

Yet when people look to a golden future they should not forget the role of hubris. Often our predictions return to haunt us. It is particularly ironic that in this same year, 1900, ideas and approaches began to surface that were to transform our world, our society, and ourselves in radical and unpredictable ways.

What were those tiny seeds that were destined to blossom in such unexpected directions? In 1900 Max Planck published his first paper on the quantum, and young Albert Einstein graduated from the Zurich Polytechnic Academy. A year later Werner Heisenberg was born. These three physicists would create the great revolutions of modern science.

In 1900 Henri Poincaré was working on an abstruse technical difficulty involving Newtonian mechanics. Over half a century later this would explode into chaos theory. Astronomers were looking forward to the opening of the great telescopes at Mount Wilson in 1904 and, in the decades that followed, Edwin Hubble would use these instruments to discover that the universe was far vaster than ever believed and, moreover, that it was continually expanding.

In 1900 biologists rediscovered the work of an obscure mid nineteenth century monk, Gregor Mendel. Ignored by the scientific community in his own day, Mendel had examined the way physical characteristics are inherited when different varieties of garden peas are crossed. Who would have guessed that exactly a century after this rediscovery of the basis of genetic inheritance, the completion of the Human Genome Project would be announced?

This same year, 1900, saw the publication of Sigmund Freud's *Interpretation of Dreams.* Much more rational than a Victorian dream book, which typically flirted with divination and the occult, it demonstrated that dreams are "the royal road to the unconscious" and, in turn, that our waking lives are ruled by the irrationality of the unconscious. That unconscious had a potential for violence and human irrationality that was to be powerfully demonstrated again and again during the twentieth century.

At the end of the nineteenth century Percival Lowell used his fortune to establish his own observatory at Flagstaff, Arizona, with the aim of discovering life on Mars. In 1900 H. G. Wells, inspired by these ideas, published *War of the Worlds,* with its image of the mass destruction of the human race. Ironically the real possibility of global destruction in the twentieth century did not arise from little green men from Mars but from human-made weapons of mass destruction.

1900 was the year when the young philosopher Bertrand Russell heard Giuseppe Peano speak at a conference in Paris. The lecture so inspired Russell that he devoted his life's work to the discovery of certainty in mathematics and philosophy. How this mathematical Holy Grail itself was eventually subverted forms the core of Chapter 2.

In 1900, inspired by the writings of John Ruskin, Marcel Proust visited Venice. He abandoned the novel on which he had been working and, determined to seek some new way of expressing "man's" confrontation with eternity, he embarked on a master plan that was to terminate in one of the major literary works of the twentieth century. It was also the year that the 18-year-old James Joyce, after having his first article published, decided to become a full-time writer. In this same year Picasso had his first exhibition and made a trip to Paris, an event that was to have a profound effect on art in the twentieth century. 1900 was also the year in which Paul Cézanne was working on his famous studies of Montagne Sainte-Victoire. The works he produced there had a revolutionary effect on painting and produced yet another form of doubt as he questioned the certainty of what he was seeing.

In the previous year Henry Ford had formed the Detroit Motor Company, which would produce the famous Model T, a car that transformed American society. Add to this Ford's discovery of mass production through the assembly line and one understands in part why, when young Henry left his father's farm, only a quarter of Americans lived in a city, yet, when he died, well over half of them were city dwellers. In 1900 there were 8,000 automobiles in the United States and 150 miles of paved road. Today the number of cars in the United States is close to 100 million.

A few years earlier, in 1896, Herman Hollerith had created the Tabulating Machine Company to speed up the processing of data using a system of punched cards. In 1911 the company's name changed to International Business Machines. The radio vacuum tube had been invented (in 1904), and so both the physical components and the business infrastructure were already in place for the creation of the computer revolution.

In the same year as the creation of Hollerith's Tabulating Machine Company, Henri Becquerel discovered the radioactivity of uranium. A few decades later, while studying Becquerel's phenomenon, the German scientist Otto Hahn realized that the atom could be split. When knowledge of this process reached the United States, colleagues persuaded Einstein to write a letter to President Roosevelt recommending the building of an atomic bomb, out of the fear that Nazi scientists would do so first. And so was born the atomic age, and with it the possibility of the annihilation of all life on earth.

While the twentieth century began with confident certainty it ended in unsettling uncertainty. Never again will we have the same degree of pride in our knowledge. In our infatuation with science and technology we overestimated our ability to manipulate and control the world around us. We forgot the power of the mind's irrational impulses. We were too proud in our intellectual achievements, too confident in our abilities, too convinced that humans would stride across the world like gods.

Today we are wiser and more cautious. We are suspicious of great plans and global promises. We view with caution the sweeping proposals of experts and politicians. We savor unbounded optimism with a generous pinch of salt.

Above all we want a better world for ourselves, our children, and our children's children. We have learned that ordinary people can have a voice. We will not put our lives blindly into the hands of politicians and institutions. We demand to be heard and we know we can be effective.

Now let us return in more detail to the twentieth century and discover the various ways in which certainty dissolved into uncertainty. Each chapter that follows tells us something about uncertainty in the worlds of art, science, economics, society, and the environment. Each adds another layer to those increasingly complex questions: Who am I? What do I know? What does it mean to be human?

FDP
Pari, Italy
2002

FROM CERTAINTY TO UNCERTAINTY

One

QUANTUM UNCERTAINTY

In 1900 Lord Kelvin spoke of the triumphs of physics and how Newton's theory of motion could be extended to embrace the phenomena of light and heat. His address went on to mention "two clouds" that obscured the "beauty and clearness" of the theory: the first involved the way light travels through space, the second was the problem of distributing energy equally among vibrating molecules. The solution Kelvin proposed, however, proved to be way off the mark. Ironically, what Kelvin had taken to be clouds on the horizon were in fact two bombshells about to create a massive explosion in twentieth century physics. Their names were relativity and quantum theory, and both theories had something to say about light.

Light, according to physicists like Kelvin, is a vibration, and like every other vibration it should be treated by Newton's laws of motion. But a vibration, physicists argued, has to be vibrating *in* something. And so physicists proposed that space is not empty but filled with a curious jelly called "the luminiferous ether." But this meant that the speed of light measured in laboratories on earth—the speed with which vibrations appear to travel through the ether—should depend on how fast and in what direction the earth is moving through the ether. Because the earth revolves around the sun this direction is always varying, and so the speed of light measured from a given direction should vary according to the time of year. Scientists therefore expected to detect a variation in the speed of light measured at various times of the year, but very accurate experiments showed that this was not the case. No matter how the earth moves with respect to the background of distant stars, the speed of light remains the same.

This mystery of the speed of light and the existence, or nonexistence, of the ether was only solved with Einstein's special theory of relativity, which showed that the speed of light is a constant, independent of how fast you or the light source is traveling.

The other cloud on Kelvin's horizon, the way in which energy is shared by vibrating molecules, was related to yet another difficult problem—the radiation emitted from a hot body. In this case the solution demanded a revolution in thinking that was just as radical as relativity theory—the quantum theory.

Bohr and Einstein

Special relativity was conceived by a single mind—that of Albert Einstein. Quantum theory, however, was the product of a group of physicists who largely worked together and acknowledged the Danish physicist Niels Bohr as their philosophical leader. As it turns out, the tensions between certainty and uncertainty that form the core of this book are nowhere better illustrated than in the positions on quantum theory taken by these two great icons of twentieth century physics, Einstein and Bohr. By following their intellectual paths we are able to discover the essence of this great rupture between certainty and uncertainty.

When the two men debated together during the early decades of the twentieth century they did so with such passion for truth that Einstein said that he felt love for Bohr. However, as the two men aged, the differences between their respective positions became insurmountable to the point where they had little to say to each other. The American physicist David Bohm related the story of Bohr's visit to Princeton after World War II. On that occasion the physicist Eugene Wigner arranged a reception for Bohr that would also be attended by Einstein. During the reception Einstein and his students stood at one end of the room and Bohr and his colleagues at the other.

How did this split come about? Why, with their shared passion for seeking truth, had the spirit of open communication broken down between the two men? The answer encapsulates much of the history of twentieth century physics and concerns the essential dislocation between certainty and uncertainty. The break between them involves one of the deepest principles of science and philosophy—the underlying nature of reality. To understand how this happened is to confront one of the great transformations in our understanding of the world, a leap far more revolutionary than anything Copernicus, Galileo, or Newton produced. To find out how this came about we must first take a tour through twentieth century physics.

Relativity

Einstein's name is popularly associated with the idea that "everything is relative." This word "relative" has today become loaded with a vast number of different associations. Sociologists, for example, speak of "cultural relativism," suggesting that what we take for "reality" is to a large extent a social construct and that other societies construct their realities in other ways. Thus, they argue, "Western science" can never be a totally objective account of the world for it is embedded within all manner of cultural assumptions. Some suggest that science is just one of the many equally valid stories a society tells itself to give authority to its structure; religion being another.

In this usage of the words "relative" and "relativism" we have come far from what Einstein originally intended. Einstein's theory certainly tells us that the world appears different to observers moving at different speeds, or who are in different gravitational fields. For example, relative to one observer lengths will contract, clocks will run at different speeds, and circular objects will appear ellipsoidal. Yet this does not mean that the world itself is purely subjective. Laws of nature underlie relative appearances, and these laws are the same for all observers no matter how fast they are moving or where they are placed in the universe. Einstein firmly believed in a totally objective reality to the world and, as we shall see, it is at this point that Einstein parts company with Bohr.

Perhaps a note of clarification should be added here since that word "relativity" covers two theories. In 1905, Einstein (in what was to become known as the special theory of relativity) dealt with the issue of how phenomena appear different to observers moving at different speeds. He also showed that there is no absolute frame of reference in the universe against which all speeds can be measured. All one can talk about is the speed of one observer when measured relative to another. Hence the term "relativity."

Three years later the mathematician Herman Minkowski addressed the 80th assembly of German National Scientists and Physicians at Cologne. His talk opened with the famous words: "Henceforth space by itself, and time by itself, are doomed to fade away into mere shadows, and only a kind of union of the two will preserve an independent reality." In other words, Einstein's special theory of relativity implied that space and time were to be unified into a new four-dimensional background called space-time.

Einstein now began to ponder how the force of gravity would enter into his scheme. The result, published in 1916, was his general theory of relativity (his earlier theory now being a special case that applies in the absence of gravitational fields). The general theory showed how matter and energy act on the structure of space-time and cause it to curve. In turn, when a body enters a region of curved space-time its speed begins to change. Place an apple in a region of space-time and it accelerates, just like an apple that falls from a tree on earth. Seen from the perspective of General Relativity the force of gravity acting on this apple is none other than the effect of a body moving through curved space-time. The curvature of this space-time is produced by the mass of the earth.

Now let us return to the issue of objectivity in a relative world.

Imagine a group of scientists here on earth, another group of scientists in a laboratory that is moving close to the speed of light, and a third group located close to a black hole. Each group observes and measures different phenomena and different appearances, yet the underlying laws they deduce about the universe will be identical in each of the three cases. For Einstein, these laws are totally independent of the state of the observer.

This is the deeper meaning of Einstein's great discovery. Behind all phenomena are laws of nature, and the form of these laws, their most elegant mathematical expression, is totally independent of any observer. Phenomena, on the other hand, are manifestations of these underlying laws but only under particular circumstances and contexts. Thus, while phenomena appear different for different observers, the theory of relativity allows scientists to translate, or transform, one phenomenon into another and thus to return to an objective account of the world. Hence, for Einstein the certainty of a single reality lies behind the multiplicity of appearance.

Relativity is a little like moving between different countries and changing money from dollars into pounds, francs, yen, or euros. Ignoring bank charges, the amount of money is exactly the same, only its physical appearance—the bank notes in green dollars or pounds, yen, euros, and so on changes. Similarly a statement made at the United Nations is simultaneously translated into many different languages. In each particular case the sound of the statement is quite different but the underlying meaning is the same. Observed phenomena could be equated to statements in different languages, but the underlying meaning that is the source of these various translations corresponds to the objective laws of nature.

This underlying reality is quite independent of any particular observer. Einstein felt that if the cosmos did not work in such a way it would simply not make any sense and he would give up doing physics. So, in spite of that word "relativity," for Einstein there was a concrete certainty about the world, and this certainty lay in the mathematical laws of nature. It is on this most fundamental point that Bohr parted company with him.

Blackbody Radiation

If Einstein stood for an objective and independent reality what was Niels Bohr's position? Bohr was an extremely subtle thinker and his writings on quantum theory are often misunderstood, even by professional physicists! To discover how his views on uncertainty and ambiguity evolved we must go back to 1900, to Kelvin's problem of how energy is distributed amongst molecules and an even more troubling, related issue, that of blackbody radiation.

A flower, a dress, or a painting is colored because it absorbs light at certain frequencies while reflecting back other frequencies. A pure black surface, however, absorbs all light that falls on it. It has no preference for one color over another or for one frequency over another. Likewise, when that black surface is warmer than its surroundings it radiates its energy away and, being black, does so at every possible frequency without preferring one frequency (or color) over another.

When physicists in the late nineteenth century used their theories to calculate how much energy is being radiated, the amount they arrived at, absurdly, was infinite. Clearly this was a mistake, but no one could discover the flaw in the underlying theory.

Earlier that century the Scottish physicist James Clerk Maxwell had pictured light in the form of waves. Physicists knew how to make calculations for waves in the ocean, sound waves in a concert hall, and the waves formed when you flick a rope that is held fixed at the other end. Waves can be of any length, with an infinite range of gradations. In the case of sound, for example, the shorter the wavelength—the distance between one crest and the next—the higher the pitch, or frequency, of the sound because the shorter the distance between wave crests, the more crests pass a particular point, such as your ear, in a given length of time. The same is true of light: long wavelengths lie toward the red end of the spectrum, whereas blue light is produced by higher frequencies and shorter wavelengths.

By analogy with sound and water waves, the waves of light radiated from a hot body were assumed to have every possible length and every possible frequency; in other words, light had an infinite number of gradations from one wavelength to the next. In this way an infinity crept into the calculation and emerged as an infinite amount of energy being radiated.

The Quantum

In 1900 Max Planck discovered the solution to this problem. He proposed that all possible frequencies and wavelengths are not permitted, because light energy is emitted only in discrete amounts called quanta. Rather than continuous radiation emerging from a hot body, there is a discontinuous, and finite, emission of a series of quanta.

With one stroke the problem of blackbody radiation was solved, and the door was opened to a whole new field that eventually became know as quantum theory. Ironically Einstein was the first scientist to apply Planck's ideas. He argued that if light energy comes in the form of little packages, or quanta, then when light falls on the surface of a metal it is like a hail of tiny bullets that knock electrons out of the metal. In fact this is exactly what is observed in the "photoelectric effect," the principle behind such technological marvels as the "magic eye." When you stand in the doorway of an elevator you interrupt a beam of light that is supposed to be hitting a photocell. This beam consists of light quanta, or photons, that knock electrons from their atoms and in this way create an electrical current that activates a relay to close the door. A person standing in the doorway interrupts this beam and so the door does not close.

The next important step in the development of quantum theory came in 1913 from the young Niels Bohr who suggested that not only light, but also the energy of atoms, is quantized. This explains why, when atoms emit or lose their energy in the form of radiation, the energy given out by a heated atom is not continuous but consists of a series of discrete frequencies that show up as discrete lines in that atom's spectrum. Along with contributions from Werner Heisenberg, Max Born, Erwin Schrödinger and several other physicists the quantum theory was set in place. And with it uncertainty entered the heart of physics.

Complementarity

Just as relativity taught that clocks can run at different rates, lengths can contract, and twins on different journeys age at different rates, so too quantum theory brought with it a number of curious and bizarre new concepts. One is called wave-particle duality. In some situations an electron can only be understood if it is behaving like a wave delocalized over all space. In other situations an electron is detected as a particle confined within a tiny region of space. But how can something be everywhere and at the same time also be located at a unique point in space?

Niels Bohr elevated duality to a universal principle he termed "complementarity." A single description "this is a wave" or "this is a particle," he argued, is never enough to exhaust the richness of a quantum system. Quantum systems demand the overlapping of several complementary descriptions that when taken together appear paradoxical and even contradictory. Quantum theory was opening the door to a new type of logic about the world.

Bohr believed that complementarity was far more general than just a description of the nature of electrons. Complementarity, he felt, was basic to human consciousness and to the way the mind works. Until the twentieth century, science had dealt in the certainties of Aristotelian logic: "A thing is either A or not-A." Now it was entering a world in which something can be "both A and not-A." Rather than creating exhaustive descriptions of the world or drawing a single map that corresponds in all its features to the external world, science was having to produce a series of maps showing different features, maps that never quite overlap.

Chance and the Irrational in Nature

If complementarity shook our naive belief in the uniqueness of scientific physical objects, certainty was to receive yet another shock in the form of the new role taken by chance. Think, for example, of Marie Curie's discovery of radium. This element is radioactive, which means that its nuclei are unstable and spontaneously break apart or "decay" into the element radon. Physicists knew that after 1,620 years only half of this original radium will be left—this is known as its half-life. After a further 1,620 years only a quarter will remain, and so on. But an individual atom's moment of decay is pure chance—it could decay in a day, or still be around after 10,000 years.

The result bears similarity to life insurance. Insurers can compute the average life expectancy of 60-year-old men who do not smoke or drink, but they have no idea when any particular 60-year-old will die. Yet there is one very significant difference. Even if a 60-year-old does not know the hour of his death, it is certain that his death will be the result of a particular cause—a heart attack, a traffic accident, or a bolt of lightning. In the case of radioactive disintegration, however, there is no cause. There is no law of nature that determines when such an event will take place. Quantum chance is absolute.

To take another example, chance rules the game of roulette. The ball hits the spinning wheel and is buffeted this way and that until it finally comes to rest on a particular number. While we can't predict the exact outcome, we do know that at every moment there is a specific cause, a mechanical impact, that knocks the ball forward. But because the system is too complex to take into account all the factors involved—the speed of the ball, the speed of the wheel, the precise angle at which the ball hits the wheel, and so on—the laws of chance dominate the game. As with life insurance, chance is another way of saying that the system is too complex for us to describe. In this case chance is a measure of our ignorance.

Things are quite different in the quantum world. Quantum chance is not a measure of ignorance but an inherent property. No amount of additional knowledge will ever allow science to predict the instant a particular atom decays because nothing is "causing" this decay, at least in the familiar sense of something being pushed, pulled, attracted, or repelled.

Chance in quantum theory is absolute and irreducible. Knowing more about the atom will never eliminate this element. Chance lies at the heart of the quantum universe. This was the first great stumbling block, the first great division between Bohr and Einstein, for the latter refused to believe that "the Good Lord plays dice with the universe."

Einstein: The Last Classical Physicist

Even now, half a century after Einstein's death, it is too soon to assess his position in science. In some ways his stature could be compared to that of Newton who, following on from Galileo, created a science that lasted for 200 years. He made such a grand theoretical synthesis that he was able to embrace the whole of the universe. Some historians of science also refer to Newton as the last magus, a man with one foot in the ideas of the middle ages and the other in rationalistic science. Newton was deeply steeped in alchemy and sought the one Catholick Matter. He had a deep faith in a single unifying principle of all that is.

Likewise Einstein, who was responsible for the scientific revolution of relativity as well as some of the first theoretical steps into quantum theory, is regarded by some as the last of the great classical physicists. As with Shakespeare, great minds such as Newton's and Einstein's appear to straddle an age, in part gazing forward into the future, in part looking back to an earlier tradition of thought.

When Einstein spoke of "the Good Lord" as not playing dice with the universe, he was referring not to a personal god but rather to "the God of Spinoza," or, as with Newton, to an overarching principle of unity that embraces all of nature. The cosmos for Einstein was a divine creation and thus it had to make sense, it had to be rational and orderly. It had to be founded upon a deep and aesthetically beautiful principle. Its underlying structure had to be satisfyingly simple and uniform. Reality, for Einstein, lay beyond our petty human wishes and desires. Reality was consistent. It reflected itself at every level. Moreover, the Good Lord had given us the ability to contemplate and understand such a reality.

Einstein could have sat down at Newton's dinner table and discussed the universe with him, something he was ultimately unable to do with Bohr. Bohr and quantum theory spoke of absolute chance. "Chance" to Einstein was a shorthand way of referring to ignorance, to a gap in a theory, to some experimental interference that had not yet been taken into account.

Wolfgang Pauli, another of the physicists who helped to develop quantum theory, put the counterargument most forcefully when he suggested that physics had to come to terms with what he called "the irrational in matter." Pauli himself had many conversations with the psychologist Carl Jung, who had discovered what Pauli termed an "objective level" to the unconscious. It is objective because this collective unconscious is universal and lies beyond any personal and individual events in a person's life. Likewise, Pauli suggested that just as mind had been discovered to have an objective level, so too would matter be found to have a subjective aspect. One feature of this was what Pauli called the "irrational" behavior of matter. Irrationality, for Pauli, included quantum chance, events that occur outside the limits of causality and rational physical law.

The gap between Pauli's irrationality of matter and Einstein's objective reality is very wide. What made this gap unbridgeable was an even more radical uncertainty—whether or not an underlying reality exists at the quantum level, whether or not there is any reality independent of an act of observation.

Heisenberg's Uncertainty Principle

This disappearance of an ultimate reality has its seed in Werner Heisenberg's famous uncertainty principle. When Heisenberg discovered quantum mechanics he noticed that his mathematical formulation dictated that certain properties, such as the speed and position of an electron, couldn't be simultaneously known for certain. This discovery was then expressed as Heisenberg's uncertainty principle.

When astronomers want to predict the path of a comet all they need to do is measure its speed and position at one instance. Given the force of gravity and Newton's laws of motion, it is a simple matter to plug speed and position into the equations and plot out the exact path of that comet for centuries to come. But when it comes to an electron, things are profoundly different. An experimenter can pin down its position, or its speed, but never both at the same time without a measure of uncertainty or ambiguity creeping in. Quantum theory dictates that no matter how refined are the measurements, the level of uncertainty can never be reduced.

How does this come about? It turns out to be a direct result of Max Planck's discovery that energy, in all its forms, is always present in discrete packets called quanta. This means a quantum cannot be split into parts. It can't be divided or shared. The quantum world is a discrete world. Either you have a quantum or you don't. You can't have half or 99 percent of a quantum.

This fact has a staggering implication when it comes to our knowledge of the atomic world. Scientists learn about the world around them by making observations and taking measurements. They ask: How bright is a star? How hot is the sun? How heavy is Newton's apple? How fast is a meteor?

Quantum Participation

Whenever a measurement is made something is recorded in some way. If no record were created, if no change had occurred, then no measurement would have been made or registered. This may not be obvious at first sight so let's do an experiment: Measure the temperature of a beaker of water. Put a thermometer in the water and register how high the mercury rises. For this to happen some of the heat of the water must have been used to heat up and expand the mercury in the thermometer. In other words, an exchange of energy between the water and the thermometer is necessary before a measurement can be said to have been recorded.

What about the position or the speed of a rocket? Electromagnetic waves are bounced off the rocket, picked up on a radar dish, and processed electronically. From the returned signals it is a simple matter to determine the rocket's position. These same signals can also be used to find out how fast the rocket is traveling—the technique is to use what is known as the Doppler shift—a slight change in frequency of the reflected signal. (This Doppler shift is the same effect you hear as a change in pitch of the siren as an ambulance or police car approaches and then speeds off into the distance.) Because the radar radiation has bounced off the rocket this means that an exchange of energy has taken place. Of course in this case the amount of energy is entirely negligible when compared with the energy of the traveling rocket.

No matter what example you think of, whenever a measurement is made some exchange of energy takes place—the rise or fall of mercury in a thermometer, a Geiger counter's clicks, the swing of a meter, electrical signals from a probe that write onto a computer's memory, the movement of a pen on a chart. In our large-scale world we don't bother about the size of the energy exchange. The amount of heat that is needed to push mercury up a thermometer is too small to be concerned with when compared to the energy of a pan of boiling water. Moreover it is always possible for measurements to be refined and any perturbing effects calculated and compensated for.

Things are quite different in the quantum world. To make a quantum observation or to register a measurement in any way, at least one quantum of energy must be exchanged between apparatus and quantum object. But because a quantum is indivisible, it cannot be split or divided. At the moment of observation we cannot know if that quantum came from the measuring apparatus or from the quantum object. During the measurement, object and apparatus are irreducibly linked.

As a measurement is being made and registered the quantum object and measuring apparatus form an indissoluble whole. The observer and the observed are one. The only way they could be separated is if we could cut a quantum into two parts—one part remaining with the measuring apparatus and the other with the quantum object. But this cannot be done. So the measuring apparatus and quantum system are wedded together by at least one quantum. What's more, the energy of this quantum is not negligible when compared with the energy of the quantum system.

This means that every time scientists try to observe the quantum world they disturb it. And because at least one quantum of energy must always be involved, there is no way in which the size of this disturbance can be reduced. Our acts of observing the universe, our attempts to gather knowledge, are no longer strictly objective because in seeking to know the universe we act to disturb it. Science prides itself on objectivity, but now Nature is telling us that we will never see a pure, pristine, and objective quantum world. In every act of observation the observing subject enters into the cosmos and disturbs it in an irreducible way.

Science is like photographing a series of close-ups with your back to the sun. No matter which way you move, your shadow always falls across the scene you photograph. No matter what you do, you can never efface yourself from the photographed scene.

The physicist John Wheeler used the metaphor of a plate glass window. For centuries science viewed the universe objectively, as if we were separated from it by a pane of plate glass. Quantum theory smashed that glass forever. We have reached in to touch the cosmos. Instead of being the objective observers of the universe we have become participators.

Heisenberg's Microscope

Our story of quantum strangeness has not yet ended. There is one further step to take—a step that Einstein could never accept and which has implications for the very nature of reality. It is a step that arose in a dispute between Bohr and Heisenberg over the interpretation of the uncertainty principle.

In the early days of quantum theory Werner Heisenberg tried to explain the origins of uncertainty much as I have done in the preceding text, by analogy with the way radar is used to ascertain the position and speed of a rocket. In the large-scale world of rockets and meteors a continuous stream of radar signals is used, but Heisenberg was thinking of an idealized sort of microscope that could be used to study an electron. This microscope would use the minimum amount of disturbance—a single photon, or quantum of light, at a time.

First, a single photon determines the speed of the electron and the result is written down. Next, a single photon determines the position of the electron and that result is written down. But by measuring this position, the electron received an impact by a photon, which changed its speed. Alternatively, in measuring the speed, the impacting photon deflects the electron from its path, thus affecting its position. In other words, Heisenberg pointed out, as soon as you try to measure position you change the electron's speed, and as soon as you try to measure speed you change the electron's position. There is always an irreducible element of uncertainty involving speed and position.[1]

It is in this way, Heisenberg argued, that uncertainty arises. It is the result of the disturbances we make when we attempt to interrogate the quantum world. Because the quantum is indivisible this uncertainty is totally unavoidable. The French physicist Bernard D'Espagnat coined the term "a veiled reality" for this property. Quantum reality by its very nature, he observed, is veiled and concealed from us. No matter how refined our experiments may be, the ultimate nature of this reality can never be fully revealed.

The Disappearance of Quantum Reality

There the matter stood until Niels Bohr stepped in. While physicists such as Werner Heisenberg, Wolfgang Pauli, Erwin Schrödinger, and Max Born were working at the mathematical formulation of the new theory, Niels Bohr was thinking about what the theory actually meant. For this reason he summoned Heisenberg to Copenhagen and confronted him about the deeper significance of his "microscope experiment."

Bohr argued that Heisenberg's explanation began by assuming the electron actually *has* a position and a speed and that the act of measuring one of these properties disturbs the other. In other words, Bohr claimed that Heisenberg was assuming the existence of a fixed underlying reality; that quantum objects possess properties—just like everyday objects in our own world—and that each act of observation interferes with one of these properties.

He went on to argue that Heisenberg's very starting point was

[1]Because a quantum is indivisible and shared between observer and observed, physics cannot say if a particular photon was produced by the apparatus, or by the observed electron, or both together. For this reason it is not possible to calculate the effect of perturbations on speed and position and thereby compensate to reduce the uncertainty.

wrong in assuming that the electron *has* intrinsic properties. To say that an electron *has* a position and *has* a speed only makes sense in our large-scale world. Indeed, concepts like causality, spatial position, speed, and path only apply in the physics of the large scale. They cannot be imported into the world of the quantum.

Bohr's argument was so forceful that he actually reduced Heisenberg to tears. Whereas Heisenberg had argued that the act of looking at the universe disturbs quantum properties, Bohr's position was far subtler. Every act of making a measurement, he said, is an act of interrogating the universe. The answer one receives to this interrogation depends on how the question is framed—that is, how the measurement is made. Rather than trying to unveil an underlying quantum property, the properties we observe are in a certain sense the product of the act of measurement itself. Ask a question one way and Nature has been framed into giving a certain answer. Pose the question in another way and the answer will be different. Rather than disturbing the universe, the answer to a quantum measurement is a form of co-creation between observer and observed.

Take, for example, the path of a rocket in the large-scale world. You observe the rocket at point A. Now look away and a moment later glance back and see it at point B. Although you were not looking at the rocket as it sped between A and B, it still makes perfect sense to assume that the rocket was actually somewhere between the two points. You assume that at each instant of time it had a well-defined position and path through space irrespective of the fact that you were not looking at it!

Things are different in the quantum world. An electron can also be observed at point A and then, later, at point B. But in the quantum case one cannot speak of it *having* a path from A to B, nor can one say that when it was not being observed it still *had* a speed and position.

Postmodern Reality

Pauli spoke of the need for physics to confront the subjective levels of matter and come to terms with irrationality in nature. It is as if physics in the early decades of the twentieth century was anticipating what has become known as postmodernism and "the death of the author."

Earlier ideas of literature held that a book or poem has an objective quality; it holds the meanings created by the author, and the reader has the responsibility to tease out these meanings during the act of reading. When at school we read a play by Shakespeare or analyzed a poem by Milton, we were told to uncover the various images, metaphors, and figures of speech that act as clues to the underlying meaning intended by the author.

The postmodern approach suggests that reading is more of a creative act in which readers create and generate meanings out of their own experience and history of reading. Likewise the author writes within the context of the whole history of literature and the multiple associations of language. Hence the author is no longer the final voice of authority, the true "onlie begetter." The reader is no longer just the passive receiver of information but the one who gives the text its life.

When Einstein spoke of the Good Lord he had in mind a notion of authorship similar to that of an earlier period; that is, of someone similar to the author of a Victorian novel. God had created the universe out of nothing and we, as its creatures, could come to understand the divine pattern of creation. Such a pattern was objective and existed independent of our thoughts, wishes, and desires. The extent to which this pattern remained veiled from us was a measure of our human limitations as readers of the divine book of creation.

Bohr and his colleagues in Copenhagen adopted a position close to that of the postmodern reader. The "properties" of the electron are not objective and independently existing, but arise in the act of observation itself. Without this act of observation, or creative "reading," the "properties" of an electron could not be said to exist as such. This was the origin of the real break between Bohr and Einstein.

Einstein had argued against the notion of absolute chance in quantum theory, although he was ultimately willing to admit that a quantum observation does disturb the universe in an unpredictable way and that the radioactive decay of a nucleus may be totally unpredictable. But he could never give up his belief that the universe has a definite existence. Even if we disturb the universe when we observe it, he believed, it still has an independent existence. Like an authorial text from the Victorian era, the universe, for Einstein, has a true, independent existence. It may be veiled from us, but nevertheless it still exists. We may not know the particular properties of an electron when we are not observing it, but such properties continue to exist. We may not know where an electron is located at the present moment, but it must have a path as it travels from A to B.

As Einstein put it, the cosmos is constructed of "independent elements of reality." Admittedly when we try to probe that reality our observations perturb things. But when we are not observing, when we are far away from a quantum system, it must have a true objective reality and it must possess well-defined properties—even if we don't happen to know what these are.

This was Einstein's sticking point. This was his most basic belief, that there is an objective reality behind the appearances of the world, even down to the quantum domain. His theory of relativity showed that, although appearances depend upon an observer's state of motion, behind these appearances stand objective laws of material reality. Provided we do not disturb the universe, it has an existence totally independent of us. He once said to his colleague, Abraham Pais, that he refused to believe that the moon ceased to exist when he was not observing it. But if Bohr were correct, then the universe, for Einstein, simply would no longer make sense.

Over the years, Einstein and Bohr met to debate this very point. Einstein would try to generate an idealized observation ("thought experiment") that would give sense to his notion of an independent reality. Bohr, in turn, would mull over Einstein's proposals and ultimately find flaws in the argument.

These "thought experiments" were never intended as actual laboratory experiments but were instead mental exercises used to discover whether some basic principle of physics was being violated. Take for example the issue of Heisenberg's uncertainty principle, which states that pairs of properties, such as momentum (speed times mass) and position, cannot both be known together with absolute certainty. A related uncertainty involves time and energy. When physicists attempt to measure the energy of a quantum system over smaller and smaller time intervals this same energy becomes more and more uncertain. For Bohr this ambiguity was basic to the quantum theory, whereas for Einstein, time and energy or position and momentum were objective realities "possessed" by the quantum theory. The only uncertainty, according to Einstein, lay in our inability or lack of ingenuity in measuring the objective properties of such systems.

When Bohr and Einstein met at the Solvay conference in 1930, Einstein presented Bohr with another thought experiment. Suppose, he said, we have a box filled with radiation and a shutter timed to open and close for a split second. The time interval is known with great precision, and in that interval a small amount of energy—a single photon—will escape from the box. Einstein now anticipated Bohr's position that the shorter the time interval, the more uncertain will be our knowledge of the amount of energy that has escaped. Einstein's special theory of relativity showed that energy and mass are equivalent, as shown by the formula $E=mc^2$. Therefore, if the box is weighed before and after the shutter opens, it will be lighter in the second weighing. This difference in mass gives a precise measure of how much energy has escaped. In this way, an accurate measure of energy is determined within a precise time interval. At this point, Einstein argued that he had demolished Bohr's claim about fundamental uncertainty.

Bohr had to be equally ingenious, and so he looked in detail at the way the box would be weighed. He posited that, if the box were mounted on a spring balance with the pointer of the balance pointing to zero, energy would escape the box at the moment the shutter opens, and in consequence, the mass of the box would decrease very slightly, and the box would move. As the box moves, so too the clock inside the box moves through the earth's gravitational field. Einstein's general theory of relativity tells us that the rate of a clock changes as it moves through a gravitational field. In this way Bohr was able to show that, because of changes in the rate of the clock, the more accurately we attempt to measure energy (via a change in the mass of the box) the greater will be the uncertainty in the time interval when the shutter is open. In this way Heisenberg's uncertainty was restored and Einstein's thought experiment was refuted.

Increasingly Einstein's objections were being frustrated by Bohr. Then, in 1931, Einstein and his colleagues Boris Podolsky and Nathan Rosen (EPR) believed they had finally come up with a foolproof example. By taking a quantum system and splitting it exactly in half (say parts A and B), and by having the two halves fly off to opposite ends of the universe, measurements made on A can have absolutely no effect on far-off B. But, because of fundamental conservation laws (the symmetry between the two identical halves) we can deduce some of the properties of B (such as spin or velocity) without ever observing it.

This paper reached Bohr "like a bolt from the blue." He set aside all his other work and repeatedly asked his close colleague Leon Rosenfeld, "What can they mean? Do you understand it?" Finally, six weeks later, Bohr had his refutation of Einstein's argument. "They do it 'smartly,'" he commented on the EPR argument, "but what counts is to do it right."[2]

By now the reader will have gathered that Bohr was an extremely subtle thinker. So subtle, in fact, that physicists still puzzle today about the implications of his ideas. In particular, his answer to the EPR experiment is still being discussed. One stumbling block was Bohr's writing style. As we have already learned, the Danish physicist was a great believer in complementarity, the principle that a single explanation cannot exhaust the richness of experience but rather complementary and even paradoxical explanations must be present. As his long-time colleague Leon Rosenfeld put it, "Whenever he had to write something down, being so anxious about complementarity, he felt that the statement contained in the first part of the sentence had to be corrected by an opposite statement at the end of the sentence."[3]

In the EPR argument, Einstein held to his belief that there must exist "independent elements of reality." He agreed with Bohr that when physicists attempt to measure a quantum system, the act of observa-

---

[2]The remarks of Bohr were made to Leon Rosenfeld. John Archibald Wheeler and Wojcieh Hubert Zurek, eds. *Quantum Theory and Measurement* (Princeton, NJ: Princeton University Press, 1983).

[3]Paul Buckley and F. David Peat, eds. *Glimpsing Reality: Ideas in Physics and the Link to Biology* (Toronto: University of Toronto Press, 1996).

tion perturbs that system. However, by observing only one part, A, of a system, when the other part, B, is located far away, no form of interaction—no mechanical force or field of influence—can possibly interfere with B.

Bohr agreed that Einstein had ruled out any mechanical influence on system B; nevertheless, he argued that "the procedure of measurement" has "an essential influence" on the very definition of the physical variables that are to be measured.[4]

With this argument Bohr felt that he had finally put an end to all objections to his "Copenhagen interpretation" of quantum theory. There were no "independent elements of reality," rather quantum theory displayed the essential wholeness of the universe. It is not a universe put together through a series of quasi-independent elements in interaction; instead what we take for elements or "parts" actually emerge out of the overall dynamics of quantum systems. Properties of a system do not exist "out there," as it were, but are defined through the various ways in which we approach and observe a system. As Bohr pointed out, the intention or disposition to make a measurement—for example, to collect the apparatus together—determines to some extent which sorts of properties can be measured. In this sense, although a "mechanical" interference between B and the apparatus used to measure A is absent, there is always an *influence,* to use Bohr's term, on those conditions that define possible outcomes and results.

One interesting contribution to emerge out of this discussion of the EPR paradox was made by John Bell who pointed out that quantum wholeness means that the two parts of the system A and B will continue to be "correlated" even when they are far apart. In no sense does A interact with B; nevertheless (and loosely speaking) B "knows" when a measurement is being performed on A. Or rather, it would be better to say that A and B remain co-related. This co-relationship has since been confirmed by very accurate laboratory experiments.

Bohr felt that his refutation spelled the final death knell to Einstein's dream of an independent reality. Einstein, for his part, was

---

[4]If the reader finds this statement difficult to understand, that particular puzzlement is shared by deep thinkers from theoretical physics and the philosophy of science.

never satisfied. The two men drifted apart to the point where deep communication between them was no longer possible. Their break symbolizes the dislocation in thought that occurred during the twentieth century, a dislocation between causality and chance, between certainty and uncertainty, objective reality and subjective reading. It is a split that remains in physics today as a form of almost schizophrenic thinking. As the physicist Basil Hiley puts it, "physicists give lip service to Bohr and deny Einstein, but most of them end up ignoring what Bohr thought and still think like Einstein."[5]

We Are All Suspended in Language

No wonder so many working physicists continue to think like Einstein, for Bohr's mind was extremely subtle. Already he had proposed that the notion of complementarity extends beyond physics into the whole of thought. Now he was questioning the very limitations of the human mind as it seeks to grasp reality.

Until the advent of quantum theory physicists had thought about the universe in terms of models, albeit mathematical ones. A model is a simplified picture of physical reality; one in which, for example, certain contingencies such as friction, air resistance, and so on have been neglected. This model reproduces within itself some essential feature of the universe. While everyday events in nature are highly contingent and depend upon all sorts of external perturbations and contexts, the idealized model aims to produce the essence of phenomena. Apples and cannon balls fly through an idealized space free from air resistance. Balls roll down a perfectly smooth slope in the absence of friction. An electrical current flows through a perfect metal, free from flaws and dislocations. Heat circulates around a perfectly insulated cycle from its source to some machine.

The theories of science are all about idealized models and, in turn, these models give pictures of reality. We shall explore this notion of

---

[5]Basil Hiley in conversation with the author.

"pictures of the world" in greater depth when we meet the work of Ludwig Wittgenstein in Chapter 4. For the moment let us examine Bohr's argument that all these pictures and models are based upon concepts that have evolved out of classical physics. Therefore they will always give rise to paradox and confusion when applied to the quantum world.

Bohr went even further. Physicists may work with measurements, mathematics, and equations but when they meet to discuss the meaning of these equations and describe the work they are doing, they have to speak using the same ordinary language (spoken or written) that we all use. Admittedly they employ a large number of technical terms and equations, but the bulk of these discussions take place in everyday language that evolved amongst human groups who live in the large-scale world and who are of a particular size and lifespan. The human scale of things is vastly different from that of atoms and electrons. As human consciousness evolved so too did notions of position, space, time, and causality. In their most basic form these concepts help us to survive and to explain the world around us. All these "large-scale" notions are so deeply ingrained within our language that it is impossible to carry on a discussion without (subtly and largely unconsciously) using them. But when we speak of the quantum world we find we are employing concepts that simply do not fit. When we discuss our models of reality we are continually importing ideas that are inappropriate and have no real meaning in the quantum domain. It is for this reason that Bohr declared, "We are suspended in language so that we don't know which is up and which is down." Our discursive thought always takes place within language, and that language predisposes us to picture the world in a certain way, a way that is incompatible with the quantum world.[6]

As soon as we ask, What is the nature of quantum reality? What is the underlying nature of the world? Is there a reality at the quantum level? we find ourselves entangled in words, pictures, images, models, and ideas from the large-scale world. The result, Bohr pointed out, is

---

[6]Wheeler and Zurek. *Op cit.*

confusion and paradox. In the end, it is better to remain silent than to create endless philosophical confusion; maybe this is why the discussions between Bohr and Einstein were doomed to end in silence. What had begun as a discussion of chance and uncertainty developed into a radical transformation of our ideas about the very nature of reality. The deep bond of affection between Einstein and Bohr was insufficient to overcome the growing split in their respective approaches to physics.

The Disappearance of Ultimate Reality

Quantum theory introduced uncertainty into physics; not an uncertainty that arises out of mere ignorance but a fundamental uncertainty about the very universe itself. Uncertainty is the price we pay for becoming participators in the universe. Ultimate knowledge may only be possible for ethereal beings who lie outside the universe and observe it from their ivory towers. But as incarnate beings, we live within the heart of the material world. We are all participators in the world, and the entrance fee we pay is living with a measure of uncertainty.

Uncertainty also exists in another and even more disturbing way, as an uncertainty about the very goal of science and philosophy. From the time of the Greeks, human beings have asked what the world is made of. They attempted to reach, through speculation and experiment, an ultimate ground or ultimate idea upon which all of reality is founded. Twentieth century scientists approached this idea of an ultimate ground by breaking matter into smaller and smaller bits and thereby discovered molecules, atoms, elementary particles, and, along with them, quantum theory.

But then Niels Bohr challenged the ability of science and the human mind to proceed further. He almost seemed to be suggesting that science as we knew it had finally reached a limit and could go no further as a means of enquiry into the nature of reality.[7]

---

[7]As a young man, David Bohm debated this issue of reality with Einstein in a series of letters. Einstein firmly held to his belief in an independent reality that is approachable through reason. In reply, Bohm argued that perhaps below our present level of knowledge there lie other levels, as yet unexpected and unexplored.

When the physicist and philosopher Bernard D'Espagnat spoke of the subatomic world as a "veiled reality" he was implying that something real must exist beyond the veil. Again Bohr cautions us against such ideas. We cannot even begin to discuss what lies beyond such a veil, or even that there is a "something" beyond the veil that could be said to have existence. Maybe, in the last analysis, there is no quantum reality. Maybe quantum reality exists only as a concept in our own minds.

And thus we are left with a mystery. Maybe there are no foundations to our world. Maybe there is no final goal toward which science can aim itself. Maybe notions of "existence" and "fundamental levels" are so ephemeral that they will vanish at our touch.

Something analogous occurred with the philosophical movement known as "the death of God," which has its roots in the writings of Nietzsche. Rather than denying the existence of God, it argued that the human construct, the "idea" of God, the human concept of the divine, had died. In its place that which remains lies beyond the limits

---

For 200 years Newton's physics was sufficient to describe the world—in case after case, it explained the phenomena of nature. It was only with more refined experiments at the end of the nineteenth century that physicists began to detect discrepancies in Newton's laws and so entered the world of quantum theory. But, as Bohm pointed out, quantum theory is really only needed when one deals with extremely small distances and time intervals or very high energies. For the rest of experience we need no more than classical (that is, Newtonian) physics. This means that our everyday world is extremely insensitive to what is going on beneath it at the atomic level, which is so effectively hidden from ordinary experience that it took 200 years of science to detect it.

But what if another level lies beneath quantum theory? It could take decades upon decades of careful science before such a hidden level is detected. And what if beneath that level there is another, and so on, in perpetuity? Maybe reality is infinite in its subtleties, and science will only be able to penetrate a small distance through its surface. Bohm's vision was of a science that goes on without limit. Yet at each step the next secret becomes harder and harder to uncover until science itself gives up in exhaustion.

Bohr argued, however, that our ability to enter into some "ultimate reality" of the quantum is doomed to ambiguity and confusion. Even Bohm's concepts of levels and ideas as fundamental and ultimate are all human-scaled images. They are based, for example, on architectural metaphors. The very moment we open our mouths to ask such questions we prejudice our investigation.

of discourse, concepts, ideas, and language. What remains is untouched and uncontaminated by human thought. It is an absolute mystery.

Is quantum theory telling us that science can only go so far in uncovering the mysteries of existence? Does it mean that at a certain point a further step will only lead to futile confusion? Quantum theory forces us to see the limits of our abilities to make images, to create metaphors, and push language to its ends. As we struggle to gaze into the limits of nature we dimly begin to discern something hidden in the dark shadows. That something consists of ourselves, our minds, our language, our intellect, and our imagination, all of which have been stretched to their limits.

Two

ON INCOMPLETENESS

In the previous chapter we saw how Nature limits the certainty we can expect from the material world and allows us to probe only so far into the mystery of reality; beyond this we are in danger of becoming lost in paradox and confusion. Does this mean that we have lost forever the hope of certainty?

If, through our acts of participation in nature, limits are placed on the extent of our knowing, then at least we should be able to find certainty in the abstract products of our own minds. Above all, shouldn't we be able to discover certainty within the world of mathematics? This is exactly what the philosopher Bertrand Russell believed as, in the year 1900, he listened to Giuseppe Peano speak with great clarity about the foundations of mathematics and decided to devote himself to proving their absolute rigor.

The Power and Beauty of Mathematics

The dream of structuring the world according to mathematical principles began long before the rise of modern science. The Pythagorean brotherhood of ancient Greece believed that "all is number." In the Middle Ages, mathematical harmonies were the key to both music and the architecture of great buildings. The paintings of Piero della Francesca (1420?–1492) take us into a world of deep mathematical order and balance. The same sense of harmony and proportion is found in the music of J. S. Bach, and, thanks to the research of the cellist Hans-Eberhard Dentler, we now know that Bach's *Art of the Fugue* was influenced by Pythagorean number symbolism.[1]

Where we find certainty and truth in mathematics we also find beauty. Great mathematics is characterized by its aesthetics. Mathematicians delight in the elegance, economy of means, and logical inevitability of proof. It is as if the great mathematical truths can be no other way. This light of logic is also reflected back to us in the underlying structures of the physical world through the mathematics of theoretical physics.

In *The Study of Mathematics,* Bertrand Russell put it this way: "[M]athematics takes us into the region of absolute necessity, to which not only the actual world, but every possible world, must conform." For the philosopher, "mathematics is an ideal world and an eternal edifice of truth. . . . [I]n the contemplation of its serene beauty man can find refuge from the world full of evil and suffering."[2] For the astronomer James Jeans (1877–1946), "God is a mathematician." And there is a saying amongst mathematicians that "God made the numbers. All the rest is made by humans."[3]

---

[1]Hans-Eberhard Dentler. L'Arte della Fuga di Johann Sebastian Bach: Un'opera pitagorica e la sua realizzazione (Milan: Skira, 2000).

[2]Frederick Copleston paraphrasing Russell, *The History of Philosophy,* vol. 8: *Bentham to Russell* (New York: Bantam, Doubleday Bell, 1985).

[3]James Jeans, *The Mysterious* Universe (Cambridge: Cambridge University Press, 1930).

Mathematics: The Ultimate Certainty?

And so we turn to mathematics for a final certainty and begin with one of the simplest and purest operations—the act of counting. Of all things, common sense tells us that counting should be totally certain and free from all ambiguity and confusion.

Let us take a telling example. In his novel *1984,* George Orwell portrayed a world in which the state controls the lives and minds of its citizens. When one of these citizens, Winston Smith, mounts a small personal rebellion he is arrested and sent to Room 101 for brainwashing. In a world in which all antisocial behavior has been eliminated the only remaining offense is that of "thought-crime." The notion of punishment does not arise in *1984.* For to punish would be to admit a flaw in the system, in that a citizen was capable of thinking and acting in ways other than those determined by the state. Instead, Winston Smith must be reeducated, and, as with a mathematical theorem, he must realize the inevitability of the state's goodness and rightness. In a world where reality is determined by Big Brother, Smith must grasp the simple fact that $2 + 2 = 5$. That is not to say he should acquiesce or simply agree to this absurd proposition. Rather, because the state wishes to welcome Smith back into its bosom, he must actually "know" and "see" that $2 + 2 = 5$. When his tormentor holds up two fingers on one hand and two on the other, for a moment at least Winston is both able to *know* and to *see* that they add up to five.

Orwell chose this corruption of the pure act of counting as a way of demonstrating the horror of a mind that had been totally controlled to the point where logic is denied and defied. Of all certainties counting seems to head the list. No matter what we may wish, no matter what society as a whole chooses to believe, counting and arithmetic remain objective certainties. We may believe that a ceremony can change the weather, we may be certain of the winner of the next race, we may be convinced that certain mental practices will change the crime rate in a city, but no matter how hard we try, we cannot "believe" that two plus two will ever equal five.

If we should ever encounter beings on other planets, beings whose lives are utterly alien to our own, of one thing we all agree: that they will also know that 2 + 2 = 4. Indeed, when human beings search for intelligent life in the universe they do so by beaming out mathematical data because scientists and astronomers are convinced that mathematics is the universal language of the cosmos.

If the substantiality of matter dissolves into uncertainty and complementarity, at least we should still find security in mathematics. This was the view held by mathematicians and philosophers at the start of the twentieth century. All that was required was a rigorous proof that mathematics *is* the ultimate certainty, a proof that is final and harbors no degree of ambiguity.

In essence, mathematicians wanted to prove two things:

1. *Mathematics is consistent:* Mathematics contains no internal contradictions. There are no slips of reason or ambiguities. No matter from what direction we approach the edifice of mathematics, it will always display the same rigor and truth.
2. *Mathematics is complete:* No mathematical truths are left hanging. Nothing needs adding to the system. Mathematicians can prove every theorem with total rigor so that nothing is excluded from the overall system

But why all the fuss? Why the need for such definitive proofs? After all, mathematics has been in existence since the time of the ancient Greeks. Great cathedrals were constructed according to mathematical principles and have stood for centuries. Mathematics sends a rocket to the moon and works out a multinational corporation's annual accounts. If mathematical answers were uncertain, or if accountants suddenly discovered that mathematics was leaving something out of a balance sheet, our financial world would come to an abrupt halt. In every case mathematics works perfectly, so why bother to dot the final "i" and cross the final "t"?

An appeal to common sense may work for most of us, but philosophers point out that, although mathematics is founded in logic, some mathematical results look bizarre and counterintuitive. We can't rely on common sense to tell us mathematics always works, they tell us; we want certainty, and we want proof of consistency and completeness.

How Do We Count?

It required the extremes of brainwashing to convince Winston Smith that two plus two equals five. But when you take a second look, is counting all that simple after all? We know how to count, but are we really sure what it *means* to count? How, for example, do you count the number of all the numbers or the number of all the fractions? Between any two integers, say 3 and 4, can be found a series of fractions—$3\frac{1}{2}$, $3\frac{3}{4}$, $3\frac{5}{8}$, $3\frac{11}{12}$, $3\frac{99}{100}$, and so on. If you think about it, it becomes clear that between 3 and 4 can be placed an infinite number of fractions. Likewise there are an infinite number of fractions between 0 and 1, an infinite number between 1 and 2, 2 and 3, and so on. Common sense tells us that, since you can put an infinite number of fractions between any two integers, the number of fractions must be vastly greater than the number of integers.

But here, mathematicians are happy to tell us, common sense is wrong. The number of all possible fractions is exactly the same as the number of all possible integers! How can this be true? To find out we'll have to further explore the world of counting. John and Jill each have a bag of candies and, as children will, they argue about who has more. However, they are so young that once they count past five candies they get confused as to what comes next. They decide to solve the problem in another way. John takes a candy out of his bag and Jill lays one of hers beside it. Then John takes another candy and Jill matches it. They continue in this way until one of the bags is empty. It turns out that, when John's bag is empty, Jill still has some candies left in her bag. Then, even though Jill cannot count, she knows that she must have more candies than John. On the other hand, if Jill's bag had emptied first, then she must have had fewer candies than John. And if both bags empty at the same time then they know they have an equal number of sweets—even though they do not know what the value of that number happens to be.

The same thing happens with the number of fractions and the number of numbers. Take a fraction and put it down on the table. Now match this with the number 1. The next fraction is matched up with 2, 3, 4, 5, and so on. Because the number of integers is infinite they will never run out. No matter how many fractions there are, the "bag" of integers will never empty and so the next fraction on the list can always be matched with an integer. In other words, the number of the integers and the number of the fractions is the same.

Does this sound like a bit of a cheat? For a layperson it may seem odd, yet mathematicians are convinced by the argument. This shows that in mathematics things are not always obvious, so it may be a good idea to take the time to prove the certainty of mathematical propositions.

What Is a Number?

Let's begin with the idea of "number" itself. We can all count. We all know that $2 + 2 = 4$. But what exactly *is* a number? How can we define it? John and Jill made an important discovery about numbers and mathematics. Jill has suddenly realized that she can do the same thing with apples as she did with candies. She can match each candy with an apple from the bowl. In this way she discovers that there are as many candies in her bag as there are apples in the bowl. She rushes around comparing everything in sight—apples and pears, candies and coins, dogs and cats, shoes and socks. In every case the method works. If she happens to have 10 candies, then even if she can't count past five she knows when she has exactly the same number of apples, candies, coins, shoes, and so on. She has realized that a sort of mental bag exists that we could call "the number ten." Into this bag can be fitted anything and everything, provided there are only 10 of them. Shoes and candies and apples are totally different things, but when there are 10 of them they have something in common and that is their number.

At the end of the nineteenth century, philosophers and mathematicians were considering precisely this issue—the definition of "number." It was the mathematician and philosopher Gottlob Frege who hit on Jill's discovery and defined "number" just as she did, in terms of classes and sets. As Bertrand Russell put it in his *Introduction to Mathematical Philosophy,* "The number of a class is the class of all those classes that are similar to it." That bit of verbiage stops us in our tracks while we try to understand what the words mean. Put another way, the number of a couple will be the class of all couples and the name of this is "the number two." Or as Russell puts it: "A number is anything which is the number of some class."[4] With his definition of "number" in terms of class, Frege felt that he had solved an important problem. Common sense had no problem with numbers, but now Frege had been able to clarify the same concept at the very foundations of mathematics.

Russell's Paradox

Then Frege heard from Bertrand Russell that there was a fly in the ointment! Frege had shown that you can put candies, apples, shoes, pigs, and so on each in their own class and match the members of one class with another and so determine what all of these different classes have in common—that is, the number of objects in each of these classes. But Russell objected: the class of all candies is not itself a candy, neither is it an apple. In other words, since the *class* of all candies is not a candy, it is not a member of itself.

There is nothing too shocking about this; it's simple common sense. Lots of classes are not members of themselves. The class of apples is not an apple; the class of shoes is not a shoe. So why not invent a whole new class called "the class of all classes that are not members of themselves"? So far so good. Now comes Russell's turn of the screw: Is this class a member of itself? or not? Trying to answer this question exposed a major problem in the foundation of mathematics and made mathematicians and philosophers worry that certainty may not be as simple or obvious as they had hoped.

Let us put Russell's paradox in the following way. Within a big library there is a room containing catalogs of books. Many of these catalogs contain references to their own titles, as well as to those of other books. But some of these catalogs do not refer to themselves. The librarian decides to make a new catalog called "The Grand Catalog that

[4]Bertrand Russell. *Introduction to Mathematical Philosophy* (Fairlawn, N.J.: Macmillan, 1955).

lists all catalogs that do not refer to themselves." He's almost finished his work when the thought strikes him, "Do I list the catalog I've just created within its own pages or not?"

"If I leave out that entry then my catalog is incomplete," he reasons, "for it has one missing entry, the title of the Grand Catalog itself." And so he begins to add the reference to the Grand Catalog. But as he does so, he realizes that he is being inconsistent because this catalog is only supposed to contain entries for catalogs that don't refer to themselves, and here he is adding a reference to the catalog within the catalog itself.

The librarian is in a double bind. If he wants to be consistent, then his catalog is incomplete. If he completes it, then it is at the expense of being inconsistent. What applies to catalogs, Russell argues, also applies to the definition of the class of "number." In one stroke Russell had demolished Frege's work and exposed something very fishy at the foundations of mathematics.

Paradoxes like this one made it even more important to establish mathematics on a firm basis in which every step is logical and every argument is transparent. As it turned out, Russell himself was one of those philosopher–mathematicians determined to undertake this program.

Principia Mathematica

Russell's interest in these questions began in that auspicious year of 1900 at the First International Congress of Philosophy held in Paris. On August 3 Russell heard the philosopher and mathematician Giuseppe Peano address the meeting. He was so impressed with Peano's clarity of mind that it marked the turning point in his intellectual career. He believed that Peano's abilities arose out of a mind that had been disciplined by the study of mathematical logic. This clarity was the key that Russell had been seeking for many years; he returned home to England and began to study Peano's work.

As he did so he recalled his school days when, while learning geometry, he had puzzled about its logical foundations. Now, with his colleague A. N. Whitehead, he embarked on a major undertaking: to discover the logical foundations of mathematics. This vast research project would result in two great volumes known as *Principia Mathematica.*

Mathematicians may have thought they were being rigorous until Russell and others pointed out that, within their arguments, mathematicians were using subtle forms of reasoning, sometimes unconsciously, that had never been properly formulated. Russell's plan was to use a formal, symbolic notation in which all rules of inference were totally explicit. It was to have:

- A system of signs
- A grammar; that is, rules for combining signs into formulae
- Transformation rules that allowed mathematicians to go from one formula to another
- Axioms
- Proofs, involving a finite sequence of formulas, starting with an axiom and proceeding step-by-step using the rules of transformation

The Notion of Proof

Russell's program involved basing mathematics on a strictly logical foundation, an idea that goes back to Euclid. The ancient Greeks had discovered a variety of facts about the geometry of the world but it was left to Euclid to gather these facts into a single consistent and logical scheme called *Elements of Geometry.*

Euclid began with definitions about the simplest possible elements of geometry—points, lines, planes, and so on. To these he added a few axioms, which are the logical starting points of his system and were so obvious, he hoped, as to be self-evidently true. For example, one of these axioms tells us that parallel lines do not meet no matter how long they are.

From the starting point of his definitions and axioms, Euclid sought to demonstrate the various theorems known to geometry, such

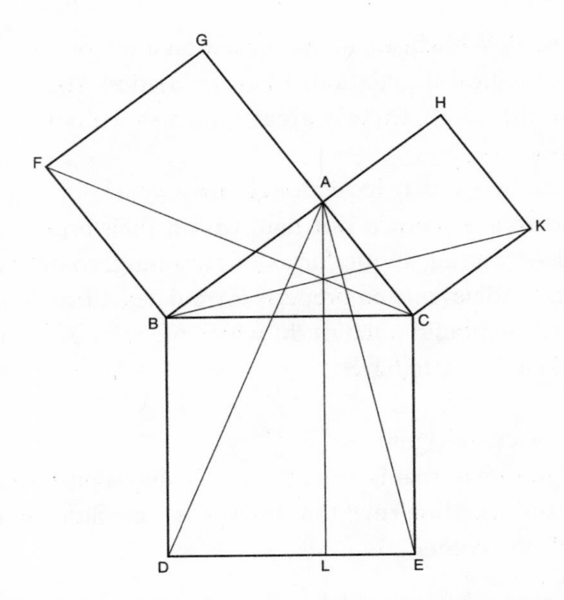

Pythagorean theorem. This theorem states that, for the right-angled triangle ABC, the area of the square BCED ("the square on the hypotenuse") is equal to the sum of the areas ABFG and AHKC ("the sum of the squares on the other two sides").

as the famous theorem of Pythagoras—the square on the hypotenuse of a right angle triangle equals the sum of the squares of the lengths of the other two sides (see figure).

At the heart of Euclid's approach lies the notion of mathematical proof. In his proofs, Euclid starts from one of the axioms, and assuming nothing else, constructs a chain of statements, each following logically onto the next. In this way it is possible to arrive at the truth of each theorem using a small number of steps and employing logic to go from one step to the next. Euclid's proofs do not involve assumptions and guesses, neither do they rely on an appeal to "common sense." Rather they are all constructed with rigorous logic.

Newton used the same approach in his great *Principles of Natural Philosophy,* first defining basic terms about space, time, and so on, and then adopting a small number of axioms as his "laws of nature." Armed with these, and proving every statement logically step by step, he was able to establish truths about the natural world.

What is particularly interesting about the theorems in Euclid's system is that, on the one hand, they were proved logically from the axioms, and on the other hand these same theorems could be tested practically with facts about the real world and the space in which we live. Euclid's method was enormously important, both because of its appeal to logic and because its theorems agreed exactly with experience. His theorems were true both within the mind and when surveyed in the field.

Mathematics Abstracts Itself

Then, in the nineteenth century, mathematicians began to ask, What happens if we change one of Euclid's axioms—just for fun? Suppose we suggest that parallel lines do meet at a point? Such a new axiom has no reference to the space in which we live. The key question was, Even with a change in one of the axioms, does the entire system still form a logically consistent, but alternative, geometry? Would this geometry be true in some alternative science fiction universe?

In short, mathematicians began to wonder about abstract axiomatic systems, systems that no longer corresponded with reality. Clearly in such totally abstract systems the issue of consistency is of paramount importance. How do we know, for example, that this alternative geometry is not free from internal contradictions?

The Power of Logic

The issue of consistency in mathematics has always been resolved by an appeal to logic. The philosopher Leibniz, for example, had argued that logic is the ideal language for philosophers. But the traditional logic of ancient Greece, Rome, and the early Middle Ages relies on purely verbal arguments: If I assume A then B must follow. Or, A thing cannot be both "A" and "not A" at the same time. Leibniz therefore proposed that verbal statements should be replaced by strings of symbols. Thus was symbolic logic born. A string of symbols says the same thing as verbal statements but in a more economical way, and, what's more, the structure of such a system is fully explicit and transparent so that it is easy to spot any error of logic. By reducing every argument to a string of logical symbols it should then be possible to analyze proofs about the foundations of mathematics in a thoroughly rigorous way.

But which proofs are to be examined? So far we've only dealt with counting, but mathematics consists of more than just numbers. What about the calculus, geometry, algebra, and so on? How, precisely, is geometry to be reduced to strings of logical symbols? To see how, let's go back to Euclid and his *Elements of Geometry*. His theorems deal with congruent triangles, bisecting circles, and so on. But Descartes showed that every point on the plane can be defined by two numbers, its x and y coordinates. Likewise, a line can be written down as an equation—the straight line $y = 3x$ or the curve $y = x^2$.

Following Descartes, geometrical figures can be represented by algebraic equations. This means that theorems in geometry can be reduced to solutions and properties of these equations. The whole of geometry, along with all its proofs, can be reduced to algebra. In turn, algebra can be reduced to theorems about numbers. And theorems about numbers can be expressed using symbolic logic. Proceeding in this fashion all of mathematics can be reduced to algebra and the rules of algebra analyzed according to symbolic logic.

So far so good. But then the mathematician David Hilbert pointed out that by reducing geometry to algebra, mathematicians had simply shifted the burden of proof to algebra. David Hilbert argued that it made more sense to make each and every aspect of mathematics formally consistent in its own right. Rather than proving geometry via algebra or interpreting points in space as numbers, each branch of mathematics should be reduced to a formal system of symbols.

Hilbert's Program

Hilbert went further by asking why we need to *interpret* geometry in terms of algebra. In a truly pure mathematics the *meaning* of these various symbols shouldn't really matter. Mathematics is simply the pure pattern of symbols, each following logically on the next according to strict rules of procedure. Rather than puzzle over the *meaning* of these symbols we should be concerned with establishing strict rules for manipulating them in order to go from one line of a proof to the next.

This was Hilbert's great program for the foundation of mathematics—his royal road to certainty. Hilbert wanted to list every possible assumption and logical principle used in mathematics: nothing was to be hidden; everything had to be up front. Rather than relying on words, every step of a proof should be replaced by rigorous strings of symbolic logic along with rules for going from one step to the next. Ideally the whole thing could be automated. Provide a computer the axioms of mathematics and a set of procedural rules, and it would work out every theorem in mathematics.

Hilbert's axiomatic approach appeared foolproof. There seemed to be no chance of making a mistake in logic. There were no hidden assumptions, nothing could exist within the system that had not previously been defined, and nothing lay outside the system other than symbolic logic. This was exactly the approach espoused by Russell and Whitehead as they worked on their vast scheme to encompass mathematics within a frame of total rigor.

Intuitionism

Not everyone agreed with Hilbert's reduction of mathematics to pure logic. The Dutch mathematician L. E. J. Brouwer argued that mathematics could not be reduced to strings of meaningless symbols alone. The notion of counting, he argued, arises out of our intuitive experience of time that allows us to distinguish the now from what is not now. It is at a deep, psychological level, he claimed, that we have the concept of "two-ness" or difference. Since our ability to count arises out of this very basic mental experience, Brouwer argued for intuitionism, an investigation of the deep psychological level at which our mathematical reasoning operates.

The Principia *Is Published*

Notwithstanding Brouwer's objections, Russell and Whitehead pushed ahead to publish their research program. The resulting text was so large that the two philosophers used a wheelbarrow to push the manuscript to the publisher's office! With the results now in print, the world's mathematicians had to decide if the two men had truly placed mathematics on a firm logical basis.

Some were still worried about Russell's paradox. Russell himself claimed that it was no more than a confusion arising out of mixing up different logical types of statement; that is, classes with classes of classes. Not everyone was convinced. Had Russell offered a true solution or was it more a matter of sweeping the problem under the carpet? What's more, some mathematicians were not happy with the standards of logical reasoning used by Russell and Whitehead.

Gödel's Theorem

Mathematicians remained undecided as to whether mathematics had been definitively established as complete and consistent. Finally, in 1931, a German paper, "On Formally Undecidable Propositions of *Principia Mathematica* and Related Systems," rocked the world of mathematics and put an end to the program of Hilbert, Russell, and Whitehead. Its author, Kurt Gödel, was 25 and living in Vienna. His paper showed once and for all that the internal consistency of the axiomatic method, sacred since the time of Euclid, is limited. More precisely, if an axiomatic system is rich enough to produce something like mathematics, then it can never be shown to be consistent. Moreover, such a system will always be inherently incomplete.

Gödel's proof was ingenious in the extreme. To begin with, he was determined to avoid the distinction between mathematics and what is known as metamathematics. In Hilbert's program, the goal was to demonstrate, using symbolic logic, that mathematics is both consistent and complete. But this meant that mathematics itself was being discussed and analyzed by another symbolic system. The system that talked about mathematics and made statements about mathematics was not itself mathematics, but metamathematics, a system that lies outside mathematics but is used to describe it.

Gödel's stroke of genius was to discover a way of remaining within mathematics by creating a symbolic system (within mathematics) that refers to itself and is therefore capable of making statements about itself—even to the point of demonstrating, or failing to demonstrate, its own consistency.

The details of Gödel's proof lie beyond the scope of this book—some hints are given in the Appendix. In essence, Gödel began by giving every symbol a number. And of course numbers very naturally fall within the province of mathematics—they are not in the field of metamathematics. By combining these numbers in a special way, he showed that every line of a proof could also be given a unique number. Every line of mathematics is defined by its own unique number. A person given that number can unpack it and write down that particular mathematical line.

Next, every theorem—along with all the lines of its proof—is also given a unique identifying number. Moreover, a statement *about* mathematics, a metastatement if you like, also has a number, and being a number it is at the same time a part of arithmetic. Gödel was finally able to arrive at numbers for statements such as "this true statement is not demonstrable," or "this statement is true" and "the negation of this statement is true." In this way he was able to show that perfectly valid numbers in arithmetic correspond to statements like "this true statement is not demonstrable." Thus Gödel was able to demonstrate that true statements exist that cannot be proved: in other words, that mathematics is incomplete.

What's more, there are numbers in his system, that is, true statements, that correspond to " this statement is true" and "the negation of this statement is true." This means that inconsistencies also exist within mathematics.

Gödel had shown that mathematics is both incomplete and inconsistent. Mathematics must be incomplete because there will always exist mathematical truths that can't be demonstrated. Truths exist in mathematics that do not follow from any axiom or theorem.

Mathematics is also inconsistent because it is possible for a statement and its negation to exist simultaneously within the same system.

Kurt Gödel's result staggered the world of mathematics. His proof appears irrefutable. The final refuge of certainty had been mathematics, and now Gödel had kicked away its last prop. But, as with something as revolutionary as Heisenberg's uncertainty principle, mathematicians and philosophers continue to ask about the deeper significance of Gödel's theorem. How is it to be interpreted? What are its implications?

To take one example, what exactly does it mean that there are true mathematical statements that cannot be proved? What would such truths look like? How would we recognize one if we saw it?

Unprovable Truths

One example of an unprovable mathematical statement may be "Goldbach's conjecture." It states that "every even number is the sum of two primes" (A "prime," or "prime number," is a number that can only be divided by itself and 1 without leaving a remainder.)

It certainly appears to work in practice, as the following examples show:

$$20 = 17 + 3$$
$$10 = 7 + 3$$
$$8 = 7 + 1$$

No mathematician has ever found an exception to this conjecture, and it has been tested on enormously large numbers using computers, though it has not been tested on every number there is—after all there are an infinite number of numbers. Mathematicians are quite certain that Goldbach's conjecture is true, but no one has ever been able to prove it. Is this the sort of unprovable truth to which Gödel was referring? Or will it turn out one day, as with Fermat's last theorem, that ingenious mathematicians will figure out a proof?

Suppose Goldbach's conjecture is a basic truth about numbers, a truth that can never be proved. Why not incorporate it as one of the underlying axioms of mathematics? All we have to do is increase the axioms of arithmetic by one and we begin a whole new ball game. Does this get us around Gödel's theorem? No, for Gödel's theorem states that once you add a new axiom, further unprovable truths will arise. No matter how you look at it, there is no avoiding Gödel's proof that mathematics is inherently incomplete.

The meaning of Gödel's result continues to be debated. For some it is a major headache, a failure to find ultimate security in logic and mathematics. Others see it in a more positive light. After all, Hilbert's great program was to reduce all mathematics to symbolic manipulations that could, in principle, be performed on a computer. A proof, Hilbert said, can be achieved through a series of algorithms, and such steps could be automated. But now Gödel is telling us that such an approach has limits and cannot encompass the whole of mathematics. There are things that human mathematicians do that can never be achieved by computers.

Limits to Algorithms

Take, for example, the idea of algorithms.[5] An algorithm is a simple rule, or elementary task, that is repeated over and over again. In this way algorithms can produce structures of astounding complexity. They can be used with a computer to produce fractals, for example. Mathematical fractals are generated by repeating the same simple steps at ever decreasing scales. In this way an apparently complex shape, containing endless detail, can be generated by the repeated application of a simple algorithm. In turn these fractals mimic some of the complex forms found in nature. After all, many organisms and colonies also grow though the repetition of elementary processes such as, for

[5]An algorithm is "a set of well defined rules for the solution of a problem in a finite number of steps" (*McGraw-Hill Dictionary of Physics and Mathematics*. New York: McGraw-Hill, 1978); or, in other words, a recipe for solving a mathematical problem.

example, branching and division. The complex pattern of tiles in a mosque is the result of a basic pattern repeated over and over again. Related patterns are also found in Arabic music. Likewise the beautiful crystal structures found in nature are the result of a repetitive process whereby atoms take up positions next to their neighbors.

Termite nests found in the tropics are several feet high and appear to be masterpieces of architectural construction. Yet no termite has in its head an overall plan of the nest. Rather, individual termites carry out extremely simple tasks of carrying particles of soil and placing them in piles. Using a simple, repetitive rule the entire nest takes shape.

There are endless examples of elaborate structures and apparently complex processes being generated through simple repetitive rules, all of which can be easily simulated on a computer. It is therefore tempting to believe that, because many complex patterns can be generated out of a simple algorithmic rule, all complexity is created in this way. Likewise, because fractals can reproduce the shapes of trees, rivers, clouds, and mountainsides it is seductive to believe that all natural systems grow and develop according to algorithmic fractal rules. Gödel's theorem points to an essential limitation in this way of thinking. A great deal of complex behavior, but not everything, can be explained through algorithms.

Take, for example, what is known as Penrose tiling. Most systems of laying down tiles—in other words of growing ever larger patterns through simple acts of repetition—require only a simple rule that shows how one tile is to be placed next to its neighbor. Proceeding in this way a person could lay down tiles all day without ever standing up to look at the overall effect. The mathematician Roger Penrose, however, pointed out that a very special system of tiling exists in which a neighborhood rule will never be sufficient to complete the pattern. Start laying down such tiles and sooner or later the next tile will fail to fit into the pattern. The only way Penrose tiles can be laid is by standing back and looking at the overall effect. Whereas algorithms work through local rules, Penrose tiles require an appreciation of the overall global plan.

What's more, certain crystals have been discovered that exhibit the same sort of symmetry as Penrose's tiles. This means that these systems do not grow simply by fitting one atom next to its neighbors; somehow the crystal *as a whole* has to have a global sense of growth. This sense of holism is exactly what one expects to find in quantum theory. A quantum system does not consist of a series of parts connected together, like a machine, but is more of an organic whole.

Cognitive Strategies

Another area in which algorithms may be found to have limits is cognitive psychology. Cognitive psychology seeks to explain human behavior and, in essence, human consciousness, through a variety of "cognitive strategies." These strategies can often be reduced to a series of algorithms that, in principle, can be simulated by a computer. It is certainly true that such strategies and algorithms appear to explain a great deal about human behavior. Likewise the related field of cognitive therapy has also been helpful to many people. The cognitive therapist identifies patterns of repetitive thinking that give rise to panic attacks, lack of self-worth, or destructive behavior within relationships. Therapy consists of making the patient aware of such patterns and using simple strategies to break the repetitive chains of thinking. But again, the implications of Gödel's results are that any system of algorithms must have inherent limitations. Maybe some aspects of consciousness and behavior can be explained through mechanistic patterns of responses, and it is certainly true that from time to time most of us do find ourselves responding in a mechanical way, yet not all of our conscious life can be explained in this way.

Artificial Intelligence

A related critique has been made regarding the artificial intelligence program. Roger Penrose, for example, argues that, although computers will become faster and more powerful, even to the point where we may no longer understand how their programs are constructed (as they began to write their own codes), they nevertheless have inherent, in-built limitations and can never achieve the degree of conscious intelligence possessed by humans.

Penrose has come under criticism from some sections of the artificial intelligence (or AI) community, yet his arguments are helpful and corrective. Again the issue is that silicon-based "intelligence" remains tied to the use of algorithms. By carrying out billions of simple repetitive tasks at very high speed, computers are able to play chess, simulate vision, recognize faces, "understand" written texts, and so on. As computers become faster, draw upon larger and larger memories, work in parallel, and employ "neural nets" that learn new tasks, they will move beyond the skills of a human being in several fields, and we may no longer understand how their "thinking processes" operate. Yet Penrose's essential point is that such devices will always be limited by Gödel's theorem, and that, by contrast, the human mind is able to make leaps and discover "truths" that can never be arrived at by stepwise logic.

In the past some quite extravagant claims have been made for the future of AI. Science fiction stories portray a world dominated by computers that outthink and outperform humans to the point where the computers finally rid the world of inefficient organic life, leaving it clean and free for machines. A more positive and more truthful vision of the future would involve a symbiosis between humans and computers. This vision acknowledges the many things computers are able to do more efficiently than humans. Their memory banks are larger. They perform calculations much faster. They don't get bored, and, provided they have been programmed correctly, they don't make mistakes.

On the other hand, these computers will be interacting with the wider society of human beings, and humans have obligations and responsibilities. We experience love, joy, heartache, and despair. We have physical bodies that interact with the world, and we possess subtleties of feeling and emotion. Human intelligence can tolerate ambiguity, make clever guesses, improvise, and patch over gaps in knowledge or logic. Human intuition can operate in highly creative ways. Humans can make leaps of logic to see patterns in disparate things. We sense what is valuable in a pattern, what is meaningful in life, and what can be safely neglected or ignored. It is in these areas that computers will encounter their limits.

It therefore makes sense to combine all that is best in these two very different species—carbon-based humans and silicon-based computers. In the future, highly advanced computers may work side by side with humans, each "learning" from the other, and each performing to the best of their abilities. Such a future may also bring direct neural connections whereby a human brain can enter directly to a computer memory, experience sensations from a remote site, or direct a robot by means of human thought and intention.

The Dominance of Logic

Gödel's theorem is about the world of mathematics, a result derived through an ingenious system of self-referential logic. To go beyond its domain of reference, as I have been doing in the preceding paragraphs, is more an extrapolation than an inference. It would be truer to say that Gödel's theorem is one example of the way in which we have learned to suspect grand, overarching schemes and ideas.

The twentieth century saw some of the darkest passages in human history; times when madness swept across entire nations and people spoke of collective evil, nightmare, and the rule of unreason. Paradoxically, such collective insanity possesses its own warped, internal reasoning; often these dark periods are characterized by an obsession with logic, bookkeeping, and bureaucracy.

Such paradoxical behavior is associated with insanity not only on the social but also on the individual level. The paranoiac carefully justifies her delusions of persecution. The psychopath, dissociated from any identification with those around him, reasons out each step, yet arrives at totally absurd premises. Psychopaths may begin with the conviction that they are superior to those around them, and in the sense that all their mental energy is focused on this obsessive idea while others seem to drift around from interest to interest and idea to idea, they may have a sort of warped justification in their belief. From that point it is a short step to seeing others as beneath them, and society's laws and conventions as applying only to such inferior creatures.

In Graham Greene's novel *The Third Man*, Harry Lime looks down on the world from atop a Ferris wheel. His is the view of the psychopath. People appear like ants, and insects are the sort of thing one crushes with one's shoe without giving them a moment's thought. Why not demonstrate one's innate superiority by destroying such an insect in an act of gratuitous murder?

The steps of reasoning are filled in, yet the conclusion is morally corrupt because healthy human beings do not entertain such thoughts. We are aware of absurdity. We are cautious of where inflated ideas may take us. We empathize with those around us and recognize another's weakness and pain.

When tied to grandiose schemes and global ideas, logic can easily sweep us away. But by arguing in this way I am not making an appeal for the abandonment of reason—that would be totally absurd. Those who formulated logic, from the time of the Greeks through the Schoolmen of the Middle Ages and on to the symbolic logic of today, have done great service to the power of human thought. On the other hand, reasoning has to be tempered with compassion, kindness, and humanity. An artist is in danger of losing sight of the whole picture if she does not stand back from the canvas to look at the wider perspective. Likewise we must constantly bracket our plans, our proposals, and our theories by asking what they mean within a broader context. How do we truly feel about them? Where are they leading us? How will others be treated by them?

When Carl Jung classified the "rational" functions of the mind he divided them into thinking and feeling. We often consider feeling to be loose and nebulous, but for Jung it was one of the mind's strictly rational functions. Feeling, for Jung, is what assesses the inherent value of things. Feeling looks at the world globally rather than analytically. If thought is not balanced by feeling, then it can become obsessive and one-tracked, giving no attention to the overall meaning of what one is doing. Conversely, if feeling is not tempered by thought, then we are in danger of rushing into events with great enthusiasm and conviction without making proper plans or understanding possible pitfalls.

If we take Gödel's theorem as a metaphor, it is telling us that something may always be left out of our grand schemes of logical thought and that inconsistencies can creep into the most logically rigorous of our frameworks. Just because things make sense on paper does not necessarily mean they will work in a practical way. Without our more human feelings logic will propel us forward, almost against our will. And when it overwhelms us it bends everything into its grip.

Quantum theory offers us an alternative viewpoint. It depends upon a logic that is inclusive and leaves room for both A and not-A. It is a logic that depends on contexts and complementarity, one in which what is A in one context becomes not-A in another. Instead of a mechanical logic that forces us onward, line by line, quantum logic invites us to step back and ask, In what context is this logic operating?

The authoritarianism of logic is a form of confrontation in which there is no middle ground. It is a logic of the excluded middle. It is a logic of winners and losers. It is a showdown in which either we triumph, so that our opponent does our bidding, or we lose face and lose power. Far better is when each voice has been heard and each position respected, when everyone has made a creative contribution and feels he or she has gained something while defeating no one. For how can "right action" flow out of anger and conflict? This is not compromise, in the sense of giving ground, but of creating a framework flexible enough to tolerate multiple points of view and contexts. It is an approach in which each person can work out of his or her own center and act in a gentle way.[6]

---

[6]The Law of the Excluded Middle. Aristotle was the key philosopher to place logic on a firm footing by showing the ways in which syllogisms (sets of logically connected steps) can be used to establish rigorous arguments. He showed, for example, that if A and B are each related to C, then it must also be true that A and B have a relationship to each other. For example:

Major Premise: All dogs (C) have four legs (A)
Minor Premise: Spot (B) is a dog (C)
Conclusion: Therefore Spot (B) has four legs (A)

But Aristotle was concerned about statements made about the future. A famous example is the proposition "there will be a sea battle tomorrow." Since, from the perspective of today, no one knows if there will be a battle tomorrow, how can this statement be treated in logic? Aristotle proposed that at least we can say the following with certainty: "It is true (now) that either there will be a sea battle tomorrow or there will not be a sea battle tomorrow." Following this line of thought Aristotle

The end of the twentieth century saw the failure of a number of grandiose schemes. We were going to green the world and discover abundant energy. To take one example out of many, in the James Bay project, extensive areas of northern Quebec were to be flooded to produce vast amounts of hydroelectric power. It was only after considerable protests stating that this virgin land supported abundant herds of caribou and was the life and culture of the Cree peoples that the most ambitious part of the project was abandoned.

Again and again such master plans proved to be insensitive to local contexts. As an antidote environmentalists adopted the slogan: Act locally, think globally. Any program should be asking: How does this relate to the world as a whole? How will it impact on each small community and ecological system?

Take, for example, the idea of a "region." Politicians draw a line on a map and call it a country, a state, a province, a county, or a region. But we can define a region in many different ways: by the accents people speak; by a network of family links; by the sort of work done; by a drainage basin, mountain range, river, or coastline; by the circulation of a newspaper; by religious groups and associations; by annual festivals; and by patterns of trade, travel, or migration. Ultimately one ends up not with a single region but with a multiplex of overlapping maps. Regions and territories depend upon a variety of contexts. But to deal with such complexity requires a more flexible and context-dependent way of thinking that is not familiar to most politicians.

As well as acting locally, we should consider the global perspective. The Amazon basin is not confined to one country. Its rainforests have an impact on the entire globe. The Rio Grande does not respect na-

---

asserted: "For any proposition *P*, either *P* or not-*P* is true." In other words, any middle or intermediate term, or proposition, is excluded and no ambiguity is present in the logical argument.

This law of the excluded middle has been much critiqued by modern logicians and a variety of alternatives have been proposed: a three-valued logic, logic that is based on laws of probability, context-dependent logic, and so on. Clearly the statement, "It is true that either the electron is a wave or it is not a wave (i.e., a particle)" does not apply at the level of quantum theory.

tional borders; neither do acid rain, ocean currents, carbon dioxide, a polluting wind, global warming, or the health of the ozone layer.

Gödel's theorem may have been a blow to mathematics, but it contains a profound lesson for us all. We had taken too much pride in the power of human reason to erect vast, impermeable towers of reason, logical systems of flawless perfection, and all-embracing canopies of knowledge. Gödel pointed out the potential flaws in this dream and showed that all-embracing schemes may contain unsuspected inconsistencies—no matter how hard we try to be comprehensive there will always be some missing knowledge. Gödel's metaphor applies to everything we do; therefore if we are to relate to the new millennium in a creative fashion we must learn new ways of thinking, ways that are more flexible and open than ever before. Rather than dealing with organizations that are rule-bound and hierarchical, we should be looking to systems that self-organize, that are organic and open in nature, and that generate their own, internal, context-dependent logics and connections.

Three

FROM OBJECT TO PROCESS

We are creatures of nature. We can't always live in a world of dreams, paradoxes, axiomatic mathematics, and uncertainties. If, from time to time, we have our head in the clouds, our feet should always be planted firmly on the ground. If we live in a high rise in the midst of a great city, we should never forget that our distant origins lay in grassy plains, rivers and streams, forests and deserts, oceans and mountains.

Our bodies are formed of matter. We require matter, in the form of air, food, and drink, in order to live. This material world is the one inalienable certainty of all life. In many of the world's religions it is symbolized by what has been called the World Tree, whose crown reaches up to heaven while its roots descend deeply toward the center of the earth. This tree is also an image of individual human life, a life that aspires to the transcendent, numinous, and spiritual by virtue of its secure foundation within the earth.

But our understanding of this stuff of the world was radically transformed by quantum theory. Chairs and tables dissolved into an empty space filled with colliding atoms. Then atoms broke apart into nuclei, nuclei into elementary particles, and finally, elementary particles into symmetries, transformations, and processes in the quantum vacuum. Understanding this new reality required a change in thinking so deep that it reached down into the very language we speak. In place of nouns and concepts we must now dialogue in terms of verbs, process, and flux. Once again, this change in our approach to reality mirrors similar revolutions that have taken place in art, literature, philosophy, and social relations.

Permanence and Change

What is the nature of this "stuff" of the world? What are the building blocks of reality? Of what substance are the foundations of all matter constructed? All cultures have grappled with this problem. It is particularly puzzling because of the apparent discrepancy between, on the one hand, the world's permanence and, on the other, its transitory nature. Compared to a human life, rocks and mountains exist forever. Set against geological ages our own lives are as contingent as the winds and weather, foods, and harvests.

Take, as an example, water. It is the most familiar and necessary of all substances. Water is always in movement and transformation. It adjusts to the shape of a vase, a cup, a swimming pool, or a dam. It falls from the sky, flows in rivers, surges in oceans, and, in a pond, its surface is rippled by the wind. On an extremely cold day this same water will freeze solid into ice; then, when the sun comes out again, this same ice melts back into water. Put some of that water into a pot over the fire and it turns into steam; place a cold spoon over the boiling water and steam condenses back into droplets of water.

Three distinct states of matter—solid, liquid, and gas—transform back and forth into each other with such perfect ease that it is natural to assume that behind these particular physical manifestations there must lie a fundamental essence common to ice, liquid water, and steam. It is as if that essence is primary, while its particular manifestation, as solid, liquid, or gas, depends on external circumstances.

What is true of water applies equally to so many other substances that surround us. Iron rusts, butter melts in the sun, meat putrefies, grape juice ferments, wine turns to vinegar, heated metals merge to form alloys. All around us are endless processes of growth and decay, and countless transformations of shapes, forms, colors, tastes, and smells. The growth of civilizations is driven by, in part, the understanding and mastery of such transformations.

Taoism of Ancient China is based on a philosophy of endless change. The worldview of the various Algonquin peoples of North America (Blackfoot, Cheyenne, Ojibway, Micmac, etc.) embraces flux and transformations. The philosophers of fifth century B.C. Greece, however, believed in an essence that lies behind such change. Thales suggested that everything is composed of water. For Anaximenes it was air. Heraclitus favored fire. Empedocles suggested a different approach: There is no single basic constituent. Rather, matter is created out of a combination of four elements—air, fire, water, and earth. Depending upon the relative proportions, substances are more earthy, fiery, airy, or watery.

Atoms or Archetypes

Associated with this idea of a fundamental foundation to the material world was the question of the divisibility of matter. Is matter continuous? Can it be endlessly divided while retaining its basic properties? Or does one finally arrive at some ultimate constituent, a basic building block that can be split no further, an *atoma* (that which cannot be divided)?

Leucippus and Democritus taught that everything is composed of elementary objects in constant movement. This proposal did not meet with the approval of Plato or Aristotle, for if everything is made of corpuscles in motion, why are the forms of things so well preserved? The atomic theory could not account for the stability of nature or for the reappearance of organic forms generation after generation. All in all, atoms appeared to be a rather mechanical explication. This certainly did not appeal to those Greek philosophers who envisioned a world of underlying forms and ideals.

All in all the Greeks preferred their *elements.* These were not actual physical substances—such as real fire or real water—but rather, non-material essences out of which the whole world was created.

Such ideas persisted in the West for well over 2,000 years, and, with the rise of alchemy, new principles, or elements, were added. The spirit Mercury, for example, is present in all that is volatile. Salt, which is unchanged by fire, represents that which is fixed, while sulfur is the principle of combustion. The Greek notion of atomism was also key in the alchemists' search for a "universal solvent" that would reduce all matter to its most elementary components.

Rather than particular substances being in their final state, alchemists believed that the components of the world are in a process of maturation and growth as they journey toward perfection. For this reason, gold was highly praised because it was considered an endpoint in alchemical workings. Gold glows like the sun and resists tarnish and dissolution. In this sense matter was a living thing, and alchemists acted as midwives to a Nature striving for perfection. The medieval doctrine of "as above so below" also established a parallelism between inner, spiritual growth and outer, material transformation.

The Rise of Atomic Theory

With the rise of "Newtonian science"—to give it an overall umbrella term—natural philosophers began to view matter in more mechanical terms, as moving in response to laws of force. Nevertheless, residues of earlier views persisted well into the nineteenth century under the guise of "vitalism," the idea that organic matter, the matter that makes up living beings, is somehow of a different order from non-organic. Such notions are still prevalent today by those who use the rather diffuse term "organic foods" to suggest that foodstuff and health products produced from "natural plants" and without the use of additives or "chemicals," have superior dietary and medicinal properties.

The first real change in the notion of elements, or fundamental components of all matter, began in the mid seventeenth century when the chemist Robert Boyle suggested that, rather than being underlying principles or forms, the elements are actual physical objects. Elements combine in different ways to form the various components of the world around us. Over a century later Antoine-Laurent Lavoisier systematically studied the many different reactions whereby substances can be broken down into their components, and the ways these components can recombine to produce a wide variety of chemical compounds. His research resulted in a list of what he believed to be the chemical elements proposed by Boyle, elements including iron, zinc, and mercury, that can never be broken down into anything simpler. For Lavoisier, these elements were the building blocks of the rest of matter.[1]

It was left to John Dalton, in the first years of the nineteenth century, to identify the notion of indivisible atoms with Lavoisier's chemical elements. Each element, he proposed, is composed of characteristic, identical atoms. These atoms link together to form molecules of various chemical compounds. What was once considered to be the result of certain basic principles, principles that also made up a person's particular character—earth, fire, air, and water—had now been transformed into little balls that interacted mechanically according to the laws formulated by Newton.

Science had uncovered a deep secret of nature, but at the expense of losing the sense of intimacy and participation that comes from believing that all nature is alive, and that we are participators under the doctrine of "as above so below." Yet, as we shall see in this chapter, the story of the nature of matter, of the movement from certainty to uncertainty, forms a great circle. The more science left the world of eternal forms and principles to voyage into the world of atoms, the more these atoms became more and more insubstantial, until matter finally vanished back into principles of form and symmetry.

Throughout the nineteenth century scientists continued to specu-

---

[1]Because of the extreme difficulty of breaking them down into more elementary components, Lavoisier believed that substances, such as silica, were also elements. Today we know that silica is a chemical compound of the elements silicon and oxygen.

late about atoms. They were used to explain the properties of gases: a gas is made up of tiny balls constantly colliding with each other. Heat the gas, and the balls move faster and for greater distances, and so the gas expands.

The first real evidence for the actual existence of atoms came in 1858 when Julius Plücker (1801–1868) noticed a curious radiation emitted as an electric current passed through a gas. Like light, these "cathode rays" traveled in straight lines but could also be deflected by a magnet. Scientists deduced that the radiation was composed of tiny electrically charged particles. In 1897 the British physicist J. J. Thomson suggested that these "electrons" are components of every atom. Thus the existence of hitherto hypothetical atoms was confirmed, and they were simultaneously found to be composite entities rather than indivisible units.

From Atoms to Elementary Particles

In 1902 Thomson and Lord Kelvin suggested that atoms are like Christmas puddings, with the negatively charged electron "raisins" embedded in a spherical "pudding" of positive charge. Then Ernest Rutherford's experiments showed that the atom is more like a miniature solar system with electrons orbiting around a central "sun" or nucleus. When two or more atoms share these orbiting electrons, atoms combine to form molecules.

But what of the nucleus itself? Physicists soon discovered that it, too, was composite and contained elementary particles called protons and neutrons. And what held this nucleus together? The Japanese physicist Hideki Yukawa proposed a new type of particle, called a meson, that binds other elementary particles together. Soon scientists discovered that not one but several different types of meson existed.

By the middle of the twentieth century there was an entire zoo of various "elementary" particles. This situation was distressing to physicists, who prefer their world to be simple and elegant. A world composed of just three types of particles would be preferable to one made up of a great many. And so the notion of the quark was proposed—

some elementary particles, such as neutrons and protons, are not elementary in themselves but are composed of various combinations of three types of quark. The theory promised to simplify the nature of matter, until scientists discovered that there had to be more than three quarks and, in addition, other sorts of particles, called gluons, to hold these quarks together.

An alternative approach was to abandon the notion of particles as fundamental building blocks in favor of superstrings, extended string-like objects whose various vibrations and rotations, quantized into a series of energy levels, produce what looks like the elementary particles. The original concept was proposed in 1970 by Yoichiro Nambu, and then revived in a striking new form by John Schwarz and Michael Green in 1984. Soon the vast majority of elementary particle physicists were working on what looked to be "The Theory of Everything."

Superstrings are incredibly tiny. Take the scale of distances between the atomic world and ourselves and double it to get down to the superstrings. What's more, superstrings don't inhabit our everyday world of three spatial dimensions but, in Schwarz and Green's theory, live in a 15-dimensional subatomic realm. For a time superstrings appeared to be the way of unifying the multiplicity of the elementary particles, but then a series of technical problems began to surface. Rather than there being a unique theory of superstrings, there turned out to be an infinite number of possible theories and there was no clear way of discerning which one was appropriate. Some physicists feel that these technical problems can be resolved (or have already been resolved) and that superstrings still hold the promise of a definitive theory of elementary matter. Others are more skeptical.

So what has happened to the Greek dream, the desire to discover the fundamental principle out of which all reality is built? What of the notion that matter cannot be divided indefinitely, but that at some point we will arrive at the fundamental building blocks of all matter?

Fundamental Symmetries

It begins to look as if the elementary particles themselves are not the final goal but rather the manifestation of underlying principles of symmetry. Just as, in Einstein's relativity, invariant laws underlie relative appearances, so too symmetry principles govern the way the elementary particles transform and group into families.

These symmetries are like mirrors so that, for example, a negatively charged electron is reflected by the mirror of charge into a positively charged positron. Likewise a proton is reflected into an anti-proton. Some mirrors reflect left-hand spinning particles into right-hand spinning ones, or reflect other properties such as hypercharge.

Of course these are not physical mirrors but rather metaphors for the way the equations that described elementary particles can be transformed and reflected one into the other. By transforming one particle into another, according to these symmetry transformations, one builds up whole families of elementary particles. In one sense these are the same particle but reflected in different ways. For many physicists the underlying laws of symmetry and transformation are more fundamental than the particles themselves.

The quantum world is in a constant process of change and transformation. On the face of it, all possible processes and transformations could take place, but nature's symmetry principles place limits on arbitrary transformation. Only those processes that do not violate certain very fundamental symmetry principles are allowed in the natural world.

Just as the ancient Greeks believed that fundamental forms and archetypes lay deeper than supposed atoms, so too contemporary physicists contrast elementary particles with more basic symmetry principles.

Grand Unified Theories

For 80 years physics has pursued the "Holy Grail" of a Grand Unified Theory, a single set of equations that is supposed to describe all that is. Like the Grail of Arthurian legend, it is occasionally glimpsed in the far distance. Yet as scientists approach more closely it disappears, or turns out to be made of tin and not gold.

One of the first of these dreams was that of Einstein, who showed that the force of gravity could be explained as the curvature of space-time. Maybe, he conjectured, magnetism and electrical attraction could also be accounted for on the same basis. Possibly matter itself is no more than knots and concentrations in the fabric of space-time. Einstein worked on this approach until the end of his life. It was a magnificent vision save for one thing: it ignored the entire quantum world. In their search for a fundamental level or principle, physicists have not been able to discover any truly satisfying way of unifying the two great discoveries of the twentieth century—relativity and quantum theory.

The best minds of three generations of physicists have struggled with the problem of unification. From time to time it looked as if a breakthrough was imminent, but then hope faded, and yet another approach was abandoned.

Postmodern Physics

The physicist Yoichiro Nambu, who developed the first string theory (precursor of "superstrings"), coined the term "postmodern physics" to express the current dilemma. Nambu suggests that the postmodern condition applies not only to literary criticism but to physics. Up to the age of atoms it was always possible to test a scientific theory directly. A theory makes certain predictions and allows calculations to be made that can be tested directly through experiments and observations. But a theory such as superstrings talks about quantum objects that exist in a multidimensional space and at incredibly short distances. Other grand unified theories would require energies close to those experienced during the creation of the universe to test their predictions.

Clearly there is no way in which these theories could ever be tested directly. Take the most powerful elementary particle accelerator known to physics and blow it up to the size of the earth, or even the solar system, and the collisions and particles it produces would still not be remotely close to events discussed in these new grand theories.

In other words, these theories are untestable in a direct way. Instead they are used to make inferences about other theories. Rather than physics producing a fundamental theory of reality that can be put to the test, it is now dealing with theories about theories, or even theories about theories about theories. It is only at the level of subtheories or sub-subtheories that theoretical predictions can be tested.

This is a dramatic change in worldview. Science always prided itself on objectivity and the ability to deal directly with nature through carefully designed experiments. But if no one can reach energies high enough to test a theory of superstrings then what is the criterion for scientific truth? Are theories to be judged, like poetry and art, on aesthetic grounds? A good poem has a unified structure, each word fits perfectly, there is nothing arbitrary about it, metaphors hold together and interlock, the sound of a word and its reflections of meaning complement each other. Likewise postmodern physics asks: How well does everything fit together in a theory? How inevitable are its arguments? Are the assumptions well founded or somewhat arbitrary? Is its overall mathematical form particularly elegant?

A New Order for Physics

The American physicist David Bohm (1917–1992) believed that this persistent failure to unify physics exposes the limits of our current way of thinking in science. What is needed is not a brilliant new idea or a novel piece of mathematics. The issue is much deeper than piecing together a unified theory of relativity and quantum theory. It involves changing our way of thinking about the physical world. As Bohm put it, what is required is a *new order* to physics.

Despite the radical differences between Newtonian physics and the world below the atom, physicists continue to make calculations using exactly the same mathematics that Newton employed—spatial coordinates and differential equations. Covering the quantum world with a coordinate grid means that, in a very fundamental sense, little has changed between Descartes and Newton on one hand, and Bohr and Heisenberg on the other. In quantum theory a coordinate grid implies that space is a backdrop against which physics is played out. Elementary particles move in space but remain distinct from it; thus a duality exists between space (or space-time) and matter. This duality goes back to Newton. Moreover, since a coordinate is a dimensionless point, space has to be continuous. But how can a continuous space be preserved right down to infinitesimal distances in a discrete quantum world?

In a truly satisfying theory both space-time and matter must emerge as the limits of something deeper. In the limits of weak gravitational fields, and speeds that are slow when compared to that of light, general relativity gives results that are indistinguishable from those of Newtonian physics. Thus general relativity can be said to embrace and include Newtonian physics as a limit. So too, a deeper theory may emerge in the future that embraces both quantum theory and relativity as its limits. Instead of attempting to unify relativity and quantum theory, in the sense of trying to bring the two theories together, these theories would emerge naturally as particular aspects of a much deeper theory.

There have been several approaches toward these deeper theories. One of these was attempted by the Oxford mathematician Roger Penrose, who began with basic quantum units he called twistors. Out of this space of twistors, he hoped, would emerge quantum theory, space-time, and general relativity. Again the theory worked only so far and the Holy Grail of unity continued to remain out of reach.

At the moment, such a deeper theory does not exist. Bohm suggested that a new order is first required. This means a radical change in the scientific language. As we saw in the previous chapter science is only paying lip service to Niels Bohr's revolutionary ideas while continuing to think in the more classical manner of Einstein. Bohm termed this classical world the "explicate order." The explicate order is our everyday world of space, time, matter, and causality. Within this explicate order, each object has its own position in space. Objects interact with each other via fields of force, or move through space to collide. This explicate order is well described in terms of coordinates and differential equations.

The quantum world is profoundly different. It is what Bohm termed the world of the "implicate order." While the explicate order deals in separateness and independence, the implicate order is holistic and mutually enfolding. The Aristotelian logic of the explicate world dictates that if A contains B then B must be inside A. But within the

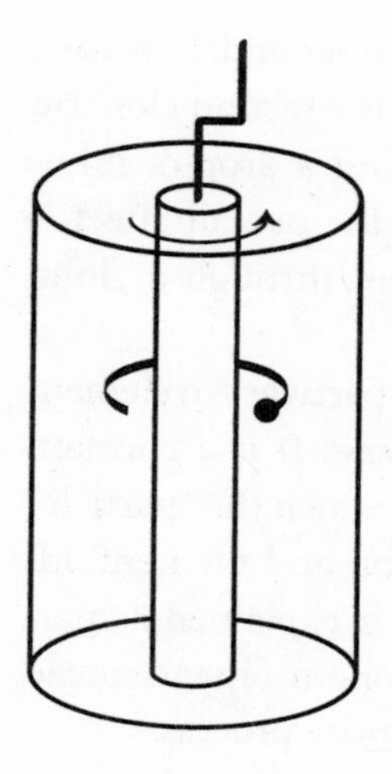

Ink drop experiment. Bohm's "ink drop experiment" gives some indication of the relationship between the implicate and explicate orders. A drop of ink is placed in glycerin and the inner cylinder rotated *n* times. As the fluid moves the drop is drawn out into an ever thinner thread until it appears to vanish, enfolded into the glycerin. When the rotation of the inner cylinder is reversed, for a further *n* turns, the drop reappears. The initial drop is analogous to the explicate order, whereas the enfolded drop is analogous to the implicate.

implicate order, A contains B at the same time that A is contained within B. Within the explicate order this would be a paradox or contradiction, but it is perfectly natural within the implicate order.

Seeking to explain this new logic Bohm supplied a few simple images that go some way toward explaining the nature of this enfolded world or implicate order. One is the ink drop experiment (see figure). Glycerin is placed between two cylinders, the inner of which can be rotated. Place a drop of ink in the glycerin, and slowly rotate the inner cylinder. The drop begins to spread out into a line. In turn, this line winds around the cylinders until it becomes so attenuated that it appears to vanish. The drop, which in the explicate order is analogous to a point in space, has become enfolded into the implicate order. Now rotate the cylinder in the reverse direction and suddenly as in a reversed movie the drop reappears, as if out of nothing. The implicate has been unfolded into the explicate.

In the next stage of the experiment, after turning the cylinder *n* times, so that the initial drop has been enfolded into the glycerin, a second drop is added close to where the first had been, and a further *n* turns are made. The process is continued with additional drops. This time not only is the first drop enfolded within the whole of the glycerin, but the second drop is enfolded within the first, and the first within the second, and so on. Now reverse the cylinder and, as before, the first drop appears but this time is followed by the second close by, then the third. Done at the right speed it looks as if a drop of ink is moving on a path through the glycerin. In fact the overall effect is rather like the way an elementary particle moves through a cloud chamber detector.[2]

This is exactly Bohm's image of an elementary particle: An elementary particle is not so much an *object* but a *process.* It is a constant process of becoming and dying away, a process in which the "particle" unfolds from the whole of space into a tiny region and then enfolds back again over all space. Wave–particle duality is explained as particular snapshots (at one moment localized, at one moment spread out) of what is not really a spatial object, but an entire process.

In terms of enfoldment from all of space, think of what happens when you look at the night sky. Light from countless numbers of stars and galaxies enters the pupil of your eye to fall on your retina. Within that tiny region of space are enfolded light and information from a vast region of the universe.

A further image of the implicate order is given by a holograph. In ordinary photography each point on a snapshot corresponds to a particular region in a scene. Here is a hand, there an eye, there a foot. There is a perfect matching of points in the scene with points on the snapshot. Holography is quite different. Each point in the scene is enfolded over the whole of the holography. Likewise, in each tiny region of the holograph can be found information about the entire scene. This means that, if a piece of the holograph is broken off and viewed, it is possible to see the entire scene and not just one fragment.

---

[2]In a cloud chamber very clean air is supersaturated with water vapor. Under normal conditions small droplets of water could condense out on dust motes to form a cloud inside the chamber. However, because the air is totally clean such condensation is not possible. But when an electrically charged elementary particle passes through the cloud chamber it hits atoms of oxygen or nitrogen (the components of air) and knocks off some of their electrons to leave electrically charged ions. Tiny droplets of water can now condense around these ions. The path of an elementary particle is recorded as a line of tiny droplets that traverse the cloud chamber. The appearance is analogous to the trail of ink drops in Bohm's double cylinder example.

These simple images, an ink drop, a holograph, and light entering the eye, do not really approach the richness of the implicate order. To lapse into "explicate language" for a moment, the implicate order is much vaster than the explicate. It is like a great ocean reaching below the surface of the explicate. Although it is always possible to unfold some aspect of the implicate into the explicate, it is never possible to expose the whole of the implicate at any one time. While concepts of larger and smaller do not really apply at the level of the implicate order, one could perhaps say, loosely speaking, that the implicate order, has the capacity to embrace and contain the explicate, but not vice versa. This means that what appear to be separate objects in our everyday world have arisen out of the same common ground and thus retain connections and attractions for each other, correlations that lie outside the normal range of explicate causality.

In order to convey some of the flavor of Bohm's ideas I have called upon images and metaphors that are somewhat static. But Bohm's notions are all about process, or the *holomovement;* that is, the movement of the whole. For Bohm, the ground (if we wish to call it that) or "all that is" takes the form of ceaseless movement. Within this movement can be discovered an endless process of unfolding and enfolding as the implicate order temporarily exposes aspects of itself to the explicate. The fact that our world appears stable is not so much that objects remain static in our world, but that the same patterns are constantly being born again only to die away as fast as thought. Our minds and bodies encounter the surface of things, and of the apparent stability of the explicate, without being truly aware of the constant movement below. (It is interesting to note that many meditative traditions lay emphasis upon the impermanence of things and suggest the world is constantly flickering in and out of existence.)

The implicate order casts light upon Bohr's complementarity. Only limited aspects of the implicate order can be made explicit, one at a time. As one unfolds into the explicate another enfolds back again and vanishes. Thus the entire implicate can never be totally accounted for. Instead, complementary aspects, such as wave and particle, are revealed one at a time, aspects that may appear, within our explicate world, to be paradoxical.

Just as Bohr believed that complementarity had relevance far beyond the confines of quantum theory so too the implicate order has a wider significance than the world of physics alone. Indeed, it turns out its most immediate appeal has been to writers and artists. Visual artists are concerned with ways of seeing and structuring the world. Beginning with Impressionism, painters began to move away from the constraints of linear, geometrical perspective in search of new orders within art. Cézanne, for example, wished to discover a new order for painting, one that would acknowledge the experiments of the Impressionists yet, at the same time, have the intellectual rigor of a Poussin. He explored form and space structured in terms of color and light, but at the same time allowed for a sense of ambiguity as, for example, a patch of green could be interpreted both as a tree in the middle distance and as foliage in the foreground.

Cézanne's paintings come close to an implicate order in the act of seeing. As with a hologram, each part of the painting is informed and enriched by every other. His portrait of the art dealer Ambroise Vollard required over 100 sittings. In the end, the painting was abandoned. Cézanne had left areas of the hands unfinished. He reasoned that, should he begin to fill in those blank areas he would in fact have to repaint the entire canvas. Thus, as Cézanne groped toward a new order for visual art, he knew that even the smallest area of the canvas was being visually enfolded into the whole.

A similar situation applies to good writing. A novel or short story contains images and metaphors, plots and subplots, protagonists and minor characters that enfold each other, each enhancing the other, so as to give structure to the overall work. John Briggs has coined the term "reflectafors" to refer to the way a metaphor can appear in a wide variety of forms throughout a work so that its inner structure is constantly reflected back onto itself. Likewise, in a piece of music the order of the entire piece can sometimes be anticipated, as enfolded within the opening bars.

Psychotherapists know that, if they are skillful enough in their interpretation, the entire course of therapy is contained, enfolded within the initial interview. The Jungian analyst Michael Conforti has referred to what he terms "the archetypal field" as being established during this first encounter—as if during this 50-minute meeting a field of attraction became structured, a field that would persist throughout the entire course of therapy that could last months or years. In turn, what transpires during therapy is so often an aspect, compressed within each therapeutic session, of the pattern of an entire life.

Indeed the Jungian archetypes themselves have something in common with Bohm's implicate order. The archetypes are the structuring principles that underlie individual and collective behavior. As structuring principles they are never perceived or experienced directly but appear as images and myths and are manifest within dreams and patterns of behavior. Someone has a dream of a person, lost in a dark wood, who encounters a white-haired man holding a map and plastic compass. The person in the dream is not an actual archetype but a particular symbolization, or manifestation, of an archetypal structuring principle. Just as one cannot encounter the implicate order directly so too one can never "see" the archetypes. Rather, one encounters their manifest forms or explicate orders. A Jungian analyst would recognize the man encountered in the wood as a particular manifestation of the archetype of the Wise Old Man and would begin to look for similar figures in the patient's dreams. Since such figures are universal to all cultures, what is of more interest are the explicate details in the dream. These have been added or created by the patient's personal unconscious. Why is the compass made of plastic and not metal? What may this be saying about the patient's relationship to his or her therapist?

From within their respective disciplines, Bohm and Jung discovered underlying and hidden orders that structure the world around us. In Jung's case, the archetypes or structuring principles of the collective unconscious can never be touched directly. They appear only through their manifestations in the consciousness and personal unconscious. In Bohm's case, one infers the implicate through its various manifestations and unfoldings into the explicate.

Archetypes and the implicate order are less theories about the world than explanatory principles. Yet Bohm also wished to develop a scientific theory appropriate to the order of the quantum world and this meant a mathematical language that would express the implicate order. Along with his colleague, Basil Hiley, Bohm studied an algebra developed in the nineteenth century by William Kingdon Clifford, William Rowan Hamilton, and Hermann Günther Grassmann. Of particular interest to Bohm and Hiley was the discovery, on looking back into Grassmann's notebooks, that this algebra had been developed as an "algebra of thought." It was a mathematician's attempt to explain the way thoughts emerge out of each other and flow in a dynamical way. The two physicists were struck by the similarities between quantum ideas and those of the processes of human consciousness. In essence it is via a mathematics of process that time enters into physics in a truly dynamical way.

A true scientific theory of the implicate order, one that could, for example, replace quantum theory, does not yet exist, although research in this field has continued after Bohm's death. In the last years of his life, Bohm was also investigating the notion of information as an actual activity within the universe. He called this "active information" and believed that a truly deep theory of nature should not fragment mind from matter.

Bohm's ideas were congenial to the neuroscientist Karl Pribram, who had been thinking along similar lines. Pribram believes that the brain is structured in ways similar to that of a holograph. One of the puzzles about brain anatomy had been the search for the "engram," the basic units whereby memories are stored in precise physical locations in the brain. On a computer's hard disc each unit of data is stored at a particular address or location. If damaged areas appear on the disc's surface, then information specifically stored in that region is lost forever. Yet when a person suffers brain damage—through a stroke, bullet wound, head injury, or the like—specific memories are not lost. Rather it is as if memory is distributed nonlocally across the entire brain.

This idea of distributed memory, along with his study of nerve connections, led Pribram to believe that the brain works analogously to a hologram, by enfolding, storing, and retrieving information from across the whole brain. This means that Bohm's implicate order universe is being perceived from within a holographic mind. Primary reality, from the atom to the brain, is of an implicate order but, for reasons of survival, we create, or project out, an explicate order world with its particular orders of causality, locality, interaction, and space and time.

We began this chapter by looking for the physical "ground" of matter. Now we find an implicate order that is much closer to the "ground of being" discussed by the philosophers of ancient Greece than to the mechanistic physics of the eighteenth century. The implicate order is not a ground in any material sense, but a constant process or "holomovement." Within this movement, inner and outer are united, mind and body, matter and mind. Out of this movement emerge specific structures and localizations in time and space that are always in the act of coming into being and dissolution.

With the rise of science a dream was born that the ultimate ground of reality would be discovered in tangible material things such as atoms, molecules, and elementary particles. It now seems that these are all manifestations of some underlying process, of symmetry principles and constant transformation.

Blackfoot Physics

Some years ago I wrote a book that explored the world through two lenses, one of Western science and the other of certain Native American groups, in particular the Blackfoot of Montana and Alberta. The Blackfoot, as do other Algonquin peoples, live in a world that seems to be very similar to that explored in the latter parts of the previous section. For them the world is flux and process. Time is constantly cycling around and nothing is fixed. In place of the permanence of objects and institutions of our world, the Blackfoot have ceremonies of renewal. By carrying out such ceremonies it is possible to renew what the flux has created, not in any fixed way, but as something closer to a vortex in a river that exists only by virtue of the water that flows through it.

While the Western world, and particularly Western science, converts the world into a series of concepts that can then be manipulated in the mind, such concepts do not come so easily in the Blackfoot language. Their philosophy deals with relationships to individual things rather than to collections of similar objects, or ideas into fixed concepts. Likewise names of things are not fixed. A person's name will change several times during his or her lifetime and to reflect particular deeds and attitudes. Neither is there a fixed concept of personality. Indeed, while we find multiple personality to be a mental aberration, the Blackfoot would view someone who believed they had only a single self, more or less fixed for life, as missing out on the richness of life's possibilities.

In place of fixed laws and organizations the Blackfoot have networks of relationships with all living things, including rocks and trees, as well as compacts that were negotiated by their ancestors with the spirits and energies of the cosmos. In a world of flux each person has an obligation to renew these relationships and compacts. And so the Blackfoot world is one of ceremony and responsibility and the recognition of life's basic impermanence. How different their vision of reality is from that which has created our vast organizations, multinationals, and government bureaucracies.

As yet the deeper meaning of quantum theory and process reality has not permeated into our general culture. However, the world of the Blackfoot does show that a society can function in a world of process, flux, and uncertainty. We shall learn a little more about such a world, and its relationship to language, in the following chapter.

We are all philosophers. At some point in our lives we have asked the deepest questions it is possible for a human being to ask. Who are we? Where do we come from? Where are we going? What is the meaning of life? Does time have an end? What is right action? What does it mean to be free? How should I act toward others? What is the meaning of death?

Since recorded history philosophers and religious teachers of all cultures have debated these questions. Some cultures have offered answers based on religion or mystical revelation. Others have created complex overarching systems of thought. Some philosophers answer these questions with yet other questions. Others seek closure and completeness and wish to create a single philosophical approach that will encompass all questions and all answers.

Some religious and philosophical systems deal in poetic images as they seek to express the transcendent. Others, particularly in the West, espouse the goals of clarity and directness. On the other hand some philosophical writings become dense and convoluted as philosophers attempt to express the ineffable in words, and force language into tasks for which it is not normally adapted.

And thus we arrive at another great question: How is it possible to say something that *means* something? How do we make sense of the world when we speak about it? How can we communicate the essence of what we feel and think about the world? How can we speak in ways that are not misunderstood? What is the correct way to use language?

The philosopher Leibniz argued that a rational and "ideal" language should only be used in philosophical arguments. Undergraduate discussions about "free will," "consciousness," "morality," and so on rapidly become bogged down in confusion over definitions. "I'm talking about one thing and you're really discussing another," we say. "Let's start by defining our terms. Let's all agree on what we're talking about." Thus the argument moves in a new direction, in trying to define free will, or awareness, or what we mean by "goodness." Yet as soon as we all agree on such a definition it seems to slip through our fingers, for we sense that we are really beginning to talk about something subtly different.

Leibniz understood these pitfalls only too well. He proposed that philosophers should adopt a language in which all terms are first properly defined and free from ambiguity. If we all agree on what is meant by "freedom," "morality," "causality," "time," "space," and so on, and we are careful only to use those terms in the way we have defined them, then we can talk together and our discussions will proceed logically, step-by-step. In this way we will arrive at a degree of certainty and freedom from ambiguity and confusion

Leibniz's program sounds ideal. Once such a language has been perfected philosophical arguments can be cleared up and, step-by-step, the great questions of philosophy resolved and answered. In this way philosophy will arrive at a general agreement of what it knows and what remains unanswered. In place of the various philosophical schools we will have total clarity. Philosophy will have placed a fence or boundary around what can be said, what can be known, and what we can say for certain. Outside this boundary will remain all the unanswered questions and degrees of uncertainty. But within the boundary the ground will be clean and free from weeds.

It was a contemporary of Leibniz, the satirist Jonathan Swift, who indicated an obvious flaw in this grand plan. His observation would be echoed three centuries later by Ludwig Wittgenstein. Swift pointed out that the dream of an ideal language is impressive, but can it exist in reality? In *Gulliver's Travels,* the protagonist visits the great academy of Lagadu where the professors of language have already eliminated everything except nouns "because in reality all things imaginable are but nouns," and, what's more, "every word we speak is in some degree a diminution of our lungs by corrosion." When in Swift's satire two philosophers wish to debate, they must avoid all ambiguity and logical inconsistency. They arrive at the debate carrying enormous sacks. Instead of using words, the first philosopher begins the debate by taking an object out of the bag and holding it up, then the other philosopher counters by holding up another object, and so on. No ambiguity or confusion is possible—a book is a book, a brick is a brick—and all the ambiguities inherent in words and language are avoided. The only problem is that the philosophers don't have much to talk about!

Swift's satire exposes the inherent weakness in Leibniz's dream. If we wish to be free from all potential confusion, we must employ a purified language, one in which all subtleties and shades of meaning have been eliminated, so that each word serves only one purpose. In this way language becomes restricted within highly narrow confines. On the other hand, if we wish to discuss the deepest issues of life, we need human language with all its richness and ability to embrace metaphor and tolerate ambiguity and paradox. This is the dilemma we shall address in the present chapter. It is a dilemma that challenges us to ask if we want to hang on to certainty or, as a basic condition of being human, accept a world that has a degree of ambiguity and uncertainty.

In Chapter 2 we learned of Bertrand Russell's search for certainty and completeness in mathematics. He was also an active player in the discussion of the nature of language. To begin with, he reacted strongly to an earlier movement in British philosophy called Idealism and, as to Hegel's grand philosophical system, Russell believed that "the whole imposing edifice of his system," as with many metaphysical systems, was founded on a logical mistake.[1] Indeed in his first significant philosophical work, *A Critical Exposition of the Philosophy of Leibniz* (1900) he argued that metaphysical arguments follow from the way language divides the world into subject and predicate.[2]

In reaction to Idealism, Russell wished to develop a philosophy of extreme clarity he called "logical atomism." His idea was to begin with the things we can know for certain about the world. In the main these are scientific statements. Russell took these as the logical building blocks or starting points for his system. He called these the "logical atoms." Just as molecules are built out of atoms, and the world around us out of molecules, so logical atomism would build a clear, coherent, and rational philosophy through a combination of logical atoms. Proceeding in this way Russell hoped to arrive at statements about the world that would be free from logical inconsistencies and confusions.

As it turned out, Russell's own student, Ludwig Wittgenstein, demonstrated the futility of this program. While language can, to some extent, be confined by the rigors of logic, through its ability to engage in metaphor, tolerate ambiguity, and embrace paradox and multiplicity, language is far happier creating jokes, making love, singing to children, exchanging gossip, praying, and writing poetry than it is in discussing the philosophy of the world. We may try to regulate and restrict language, but as soon as we begin to talk together language escapes from our control and goes its own way. In rejecting Russell's grand program for language, Wittgenstein set up his own program for philosophy, one which has had an enormous influence on thinking right down to the present day. In essence, by demonstrating the many ways in which language functions and plays with the world, Wittgenstein established what could best be described as a form of therapy for philosophers to help them out of the various dilemmas they had created for themselves.

---

[1]Bertrand Russell. *History of Western Philosophy* (London: Allen and Unwin, 1961).

[2]Bertrand Russell. *A Critical Exposition of the Philosophy of Leibniz* (New York: Cambridge University Press, 1951).

Wittgenstein's personality and approach are so exceptional that it is worth spending some moments to unfold his life story, for the creations of any individual can never be divorced from his or her personal life. Wittgenstein was born into a wealthy and cultured family in the Vienna of Gustav Mahler, Sigmund Freud, Arnold Schoenberg, the stories of Arthur Schnitzler, and the architecture of Adolph Loos. Johannes Brahms was a regular visitor to the Wittgenstein home, and Ravel dedicated his *Piano Concerto for the Left Hand* to Ludwig's brother Paul, who had been wounded in World War I.

Wittgenstein studied at home until the age of 14 when he went to school at Linz. His original plan, to study physics with Ludwig Boltzmann, was thwarted by that scientist's suicide.[3] Instead of doing physics Wittgenstein enrolled in Manchester University and in 1908 began to study aeronautics. His attempts to design a new type of propeller involved a great deal of mathematics, and his interest gradually shifted to the foundations of that subject. At that time Bertrand Russell's *Principles of Mathematics* (a prelude to the great *Principia Mathematica*) had been published, and so Wittgenstein began to read Russell and Frege. As a result he moved to Cambridge in 1911 and rapidly learned all Russell had to teach him.

To Russell, Wittgenstein appeared a young man of incredible intellectual brilliance, but also deeply tormented and at times driven to thoughts of suicide. After two years with Russell, Wittgenstein moved to a farm at Skjolden, northeast of Bergen, Norway. There he remained, thinking deeply about philosophical problems, until the outbreak of World War I, when he volunteered for the Austrian army.

Already Wittgenstein had become deeply critical of Russell's logical atomism. Russell had made significant contributions to the foundations of mathematics and was also well known to the general public. But as a serious philosopher his reputation was less secure. Russell's skill lay in the ease and clarity of his writing, and he was not afraid to

[3]Boltzmann had invented statistical mechanics, founding the science of thermodynamics upon the motion of underlying molecules. The severity of Ernst Mach's attack on his theory was one of the motives for Boltzmann's suicide.

popularize philosophy for the consumption of the general public—the kiss of death for many an academic in the Anglo-Saxon world at least. To his critics, this clarity of mind had a subtle drawback. Russell could plunge to the heart of an argument with great confidence and present its bare bones, but in so doing he was at risk of glossing over the inner subtleties of an issue.

Wittgenstein, by contrast, would not let an argument sit; he would mull over it, revisit it, and tease out its subtleties. While Russell forged ahead with his logical atoms, Wittgenstein, in his self-imposed exile, wondered: How can I say anything? How can an utterance mean anything? What is the relationship of language to the world? What can be said, or known, and what cannot be said? He wrote his deliberations in notebooks that he carried with him to the Eastern and Italian fronts.

In 1914 Wittgenstein had a revelation about the nature of language. The story goes that he was reading about a court case involving a traffic accident. To illustrate the case the court had been shown a model of the incident using miniature cars, roads, and houses.[4] It struck Wittgenstein that the reason this model worked, and the way it represented a possible state of affairs in the world, was because each of the elements in the model—a car, a road, a house—corresponded to, or pointed to, something in the real world. It was not simply the correspondence between toy cars and real cars that struck him, but something more general. It was that the arrangement of the toy cars and houses corresponded to an arrangement of objects and events in the real world.

Wittgenstein immediately pounced on the idea that language works because it presents a *picture* of reality. If you make a statement such as "the cat is chasing the mouse" each of the words corresponds to an object in the world. But more than this, the arrangement of words in the statement corresponds to a particular state of affairs in the world. This is why, Wittgenstein argued, language has meaning and allows us to say things about the world.

---

[4]Other versions of this story refer to Wittgenstein having seen a map of the accident in a newspaper.

Wittgenstein spent the end of the war as a prisoner of the Italians but was lucky enough to have his "Logical-Philosophical notebook" in his rucksack at the time of his capture. He sent this to Russell, who wrote the introduction and arranged to have it published. The philosopher G. E. Moore gave it the rather pompous title of *Tractatus Logico-Philosophicus.*

In fact Wittgenstein objected so strongly to Russell's introduction and what he felt to be Russell's misinterpretation of the book that he washed his hands of its publication. Indeed, this was to be a recurring theme in Wittgenstein's life, that he was constantly being misunderstood and misread, even by his own students. He had little hope that things would be any better in the future and felt he was writing for people with other sorts of minds.

Wittgenstein's notebook must be one of the shortest works in philosophy, yet it is one of the most significant of the twentieth century. In just 75 pages of short numbered statements Wittgenstein delineated what can be said from what cannot be said and must be passed over in silence.

Wittgenstein's propositions set out his picture correspondence between language and the world. The book begins with the first proposition: "The world is everything that is the case." It continues using propositions, subpropositions, and sub-subpropositions, each with its appropriate number and subnumber, to set down everything that can be said in a precise way.

As with Russell's logical atoms, Wittgenstein's propositions, the things that can be said clearly about the world, are close to scientific statements. According to him, these are the only sorts of things we can say about the world. On the other hand we human beings don't normally speak in scientific statements. We want to talk about our hopes, desires, and fears. We want to know what the world means, if it has a purpose, and how all this relates to the values of our own life.

For Wittgenstein these nonscientific statements cannot be stated clearly and in such a way as to picture some corresponding state in the world. Thus, he says, "the sense of the world must lie outside the world." The *Tractatus* begins "the world *is* the case" and everything in the world *is* in the world, and what happens in the world happens. But to ask about the values and meaning of things is to be concerned with something exterior to the universe. And so, for Wittgenstein, the meaning of the universe is not a fact *within* the universe.

This means that most of philosophy—ethics, the nature of freedom, the role of consciousness, and so on—cannot be formulated in the form of propositions that can be judged as true or false. Take, for example, death, which faces us all. Wittgenstein says, "Death is not an event in life. Death is not lived through." And thus Wittgenstein admonishes philosophy to "say nothing except that which can be said."

But what of that great philosophical tradition that goes back to the ancient Greeks: the search for truth? The true business of philosophers, Wittgenstein argues, is not to make such grand statements about the world but to clear up logical confusions that arise because of the way language works.

To take a crude example: I can say "the boiling snow" or "the square circle" without violating the rules of grammar. English allows me to say such things, even if they don't make sense. According to Wittgenstein the great debates in philosophy (about free will, consciousness, the origins of morals, causality, and the categories of space and time) all end up involving similar language confusions. The business of philosophy is not to seek answers to these questions but to be on a constant alert to linguistic confusions and then to clear them up.

It is as if Wittgenstein had put a boundary around language and said, "anything within this fenced-off area belongs to philosophy, all that is outside becomes the province of mystics, poets, and lovers. These latter are not trying to make 'pictures' of reality but are professing something profoundly different."

And what if we ask that burning question about the meaning of it all? Here Wittgenstein touches on mysticism. The great mystery is not "*how* the world is," but "*that* it is," he says. And as to questions of eternal life, isn't it true that our present life, the time we spend here on earth, is every bit as mysterious as any speculation about eternal life?

But suppose the layperson will not accept this. Supposing he or she demands more of the philosopher: "You've got a nice comfortable job in a university. You don't have to do much more than sit around and think. So give us some answers and don't keep on pussyfooting about language." To this Wittgenstein replies that the honest philosopher has an obligation to demonstrate that such deep questions do not in themselves have any strict meaning. And, yes, Wittgenstein agrees, the layperson is right to be critical, maybe there *is* no point in doing philosophy any more, other than in attempting to clear up confusion. Maybe there is really nothing more for a philosopher to say—the rest must be left to a Shakespeare or a Goethe. Maybe it is time for philosophers to resign their chairs and take up more useful occupations. After all, Spinoza made a living grinding lenses!

Like a Zen master, Wittgenstein leads philosophy to the brink, to the point of its nonbeing. Yet in the end, we can object to his method by pointing out that Wittgenstein is no more than a confidence trickster. If all that can be said for certain are the statements of science, then how did we end up with the *Tractatus* and its statements about riddles and the limits of language? Where did all that come from?

Wittgenstein agrees with our objection. If anyone has truly understood him, they will realize that what he has said is really "senseless." His words have been no more than a ladder used to reach a certain point. The reader who truly understands must "throw away the ladder, after he has climbed up on it." For when he truly *sees* the world rightly he can dispense entirely with Wittgenstein's propositions.

And so the *Tractatus* ends: "Whereof one cannot speak, thereof one must be silent." Wittgenstein had achieved certainty in what could be said, but at a very great price. All his life he struggled to remain honest to himself and to his philosophy. Moreover, he took his own advice, and having sent his manuscript to Russell, retired from philosophy, though he did, from time to time, meet with philosophers who wished to talk to him during this period.

Wittgenstein now began to study the religious and ethical writings of Tolstoy and reread the gospels. On being released from the prisoner-of-war camp he gave away the considerable fortune he had inherited from his father and took a job as an elementary teacher in a series of small Austrian villages.

By 1925 frictions with the other teachers and villagers caused him to resign. He thought of entering a monastic order and for a time worked as a gardener's assistant. In 1926 he designed and built a mansion in Vienna for one of his sisters and if he had continued along this path he could well have made a successful career as an architect.

Then in 1929, at the age of 40, Wittgenstein decided to return to philosophy and the University of Cambridge. The trigger may have been a lecture he had heard the previous year in Vienna, given by L. E. J. Brouwer, on the foundations of mathematics. Ironically, because he had not formally completed his Ph.D., this major philosopher (or anti-philosopher) was forced to register as a graduate student. A year later, however, he was made a fellow of Trinity College, Cambridge.

Wittgenstein returned to philosophy because he realized that there was more to be said about language. He did not, however, seek to publish any major book, make a grand summing-up, or create an overarching philosophical system. His remaining years as a philosopher were spent lecturing and talking to students. As to academic life itself, he thought little of it and refused to dine at High Table. One story has it that he set up his own card table in the dining room so that he could eat without having to talk to other academics. Instead of teaching in a lecture hall he preferred his own sparsely furnished room where students would bring in chairs and cushions.

Wittgenstein did not lecture on known topics, or explain established philosophical principles. He simply talked without notes and thought out loud in front of his students. Wittgenstein was doing philosophical research on the fly and constantly arriving at new results. Sometimes he would berate himself for being slow and stupid, other times he would simply wait in silence. At still other times there would be a lively conversation. With great concentration he would bring the group to a question that his students were supposed to answer. This would lead, in turn, to other questions. When he was dissatisfied and depressed with his lectures he would ask a student to accompany him to a film where he insisted on sitting in the first row so that he could be utterly absorbed.

One of the pathways he was exploring was the limitation of his earlier picture theory of language. Wittgenstein related many anecdotes about this to his students and friends and they varied from version to version. One story has to do with his explaining to an Italian economist, P. Sraffa, that a proposition in language must have the same logical form as the events it describes in the world. Wittgenstein claimed that there was always a particular grammar to such a proposition. In reply Sraffa made the familiar Italian gesture of dismissal or contempt, sweeping his fingers and back of the hand outward from under the chin and asking, "What is the form of that?"

Wittgenstein was struck that, while the gesture had a very clear meaning, it did not correspond to anything in the world. Now, in his encounters with his students, he began to explore the richness and complexity of language. He showed that meaning has less to do with picturing reality than with knowing about the different ways language is used and the various ways it works.

In the *Tractatus* he had erected a barrier around what could be clearly said. Now he realized that he had restricted language and interfered with its freedom. Nevertheless, part of his original argument remained valid: instead of attempting to arrive at universal truths, philosophy should be pointing out nonsense, resolving confusions, and being ever clear about what language is doing. Philosophy, he continued to assert, "will never reach the essence of truth about the world." He wasn't at all sure if there existed some sort of hidden meaning that would tell us the true nature of "mind," or "justice," or "God." Just as Niels Bohr had questioned whether a "reality" does exist below the atom, Wittgenstein questioned whether certain philosophical "truths" could be said to have an existence.

In investigating the many ways we use language Wittgenstein liked to pick the example of a game. Suppose a being arrives from the planet Mars and asks: "What is a game?" I switch on the television and show a football match, and a baseball game. "Ah!" the Martian realizes, "then a debate in the British House of Commons must be a game because it has two teams, a set of rules, and one team wins and the other loses."

In reply I point to children playing in the street and hand the Martian a book on chess and say, "these are also games." The Martian is naturally confused and pushes me to define exactly what is a game. Does it have to have sets of rules and precise strategies like chess? Or must there always be two teams like baseball? And if all-in wrestling is a game then what about ballroom dancing? Does every game involve competition of one person or team against another? Then what about solitaire? And if solitaire is a game in which there are no other participants, then is a crossword puzzle a game? And what about mathematics homework? And are all those people on the floor of the stock exchange playing a game?

I keep pointing to different games, as well as telling my visitor that a debate and a planning meeting are not games. "But," the Martian argues, "there must be some essence of a game. There must be a yardstick against which to measure things and say 'that is a game and this is not.' Otherwise why are you so confident that some things are games and others are not? How on earth do you know?"

The notion of some sort of essence to a game goes back to Plato and his Ideas. Plato said there is an Idea of a chair, the perfect form of a chair, and that real chairs are just copies of the Idea. If we didn't have this Idea in our minds how would we ever recognize a chair when we saw one? Does this mean there is an Idea of a game, to which all games participate more or less?

Nonsense and philosophical confusion, says Wittgenstein. Just because we give something a name does not mean that this corresponds to a single defining class. There is no great game in heaven to which all earthly games should conform in some way. Talking about games helps to illustrate the way language works and the sorts of confusions it can engender if we are not careful.

There is no blanket definition of a game, no well-defined class into which all games will neatly fit so that everything outside that class is clearly not a game. Nevertheless, we have no problems in talking about games and in sorting them out from activities that are not games. Language can handle that with ease.

Wittgenstein suggested that, in the case of games, things work through what he called "family resemblances." Chess and checkers resemble each other. Both are board games but they also have something in common with football—two teams advancing and attacking. Here the family resemblance is shared with rugby and field hockey, all of which use balls. Field hockey is also close to ice hockey, which is not played on grass and doesn't use a ball. These field games have something in common with volleyball—two teams and a ball. And volleyball bears a family resemblance to tennis and badminton—they also involve balls, bats, and a net. From there we can move to squash that has no net, but still involves hitting a ball with a bat. In this way, through a series of relationships, one arrives at a whole network of games without ever needing an exhaustive definition of "game" or invoking "the class of all games."

What is true about the idea of a game is equally true about "truth," "beauty," "freedom," "mind," "consciousness," and "God." Trying to define these terms or pin them down only leads us into endless difficulties because it interferes with the essential freedom and creativity of language. If you want to know what a term means, Wittgenstein suggested, then look at what it does. Look at the various ways it is used in language.

On other occasions Wittgenstein compared a word to levers in the cab of an engine. In one sense they are all levers. Yet each lever does something different. To know all about levers it is necessary to see how the different levers are used.

Problems in philosophy, Wittgenstein suggested, arise when two or more people employ the same word but use it in subtly different ways. If they both use the word "freedom" or "consciousness" this does not mean that they are necessarily talking about the same thing. Each will be using the word in different ways and linking it to different aspects of that word's entire "family resemblances." On the other hand, if they begin by defining the word, then other problems arise because the way language works means that the word in question is always slipping away from its definition as it is being used in different ways.

Language simply cannot be restrained and restricted. But this doesn't mean that we should not be very careful about what we are saying and pay great attention to the ways language is being used in different situations.

Wittgenstein continued to investigate a host of problems involving the way we talk and the different ways in which we can mean something. For example, he looked at the way we talk about colors, and asked what it would mean if a dog could speak.

Of philosophy Wittgenstein once said that it is as if a man finds himself trapped in a room. In vain he attempts to exit via the window or up the chimney. It is only when he turns around that he realizes that the door has been open all the time.

In 1947 Wittgenstein resigned his chair at Cambridge, his only source of income, to spend his remaining years at a guesthouse near Dublin and at a cottage in Galway. It was only the need for medical treatment for cancer that caused him to return to England, where he died in 1951.

Wittgenstein never published any major works after his return to Cambridge; neither did he build a great philosophical structure or come to any grand conclusions that could be taught in a course on philosophy. His approach has been called a psychotherapy of philosophy, for it offers a way to unravel philosophical confusions, or as Wittgenstein put it, a way to let the fly out of the bottle. His philosophical contributions, following the *Tractatus,* were gleaned from the notes taken from his lectures and dictations or from notebooks he kept. It was only after his death in 1951 that these various works were collected and published to form a remarkable second phase of work.

Bohr and Language

In many ways Niels Bohr complements Wittgenstein, particularly with his remark that we are suspended in language in such a way that we do not know which way is up and which is down. Wittgenstein, in his early years, argued that philosophy can only speak clearly about what is "in" the world. It should avoid making statements that are "about" the world, such as its meaning, or the nature of life and death. Bohr, for his part, restricted the limits of this "world." We human beings are creatures of a certain size and with lives of a particular duration and so our language evolved in such a way that it reflects these conditions. We are so deeply embedded within language that we may not even recognize that we are using concepts about space, time, and causality that belong to our large-scale world. This is a world in which quantum features have already been averaged out. It is when we try to talk about the quantum world, and apply our models and ideas, that confusion arises, for our very tools of communication are inappropriate to such a world.

Wittgenstein's early "picture theory" of language suggests that we are able to say things because language points to things in the real world. In Bohr's case there is nothing within language that will point to an underlying quantum world. Of course, Wittgenstein modified his position to view language in a more flexible light and yet Bohr's structure on what we can say about the quantum world still appears to hold.

The Blackfoot and the Rheomode

But are all languages the same? Or do some provide a linguistic point of entry into a quantum world of constant flux and transformation? In the previous chapter we met a society that lives in a world of constant flux. Does this mean that their language is also different from ours, and maybe even more adapted to a discussion of quantum theory? Blackfoot elders say that the way they talk amongst themselves is profoundly different from the way they talk to non-Native people, and that associated with this is a very different way of thinking about picturing the world.

Before we examine this claim we must first look at a somewhat controversial theory called the Whorf–Sapir hypothesis. This states that the language spoken by a particular society is deeply connected to its worldview. The way societies structure events and history and understand time and spatial separation can be discovered by carefully examining the way their languages work.

The debate surrounding the Whorf–Sapir hypothesis tends to obscure its significant contribution as a way to approach alternative worldviews. The currently fashionable Chomskian approach to linguistics argues that the differences between the world's language are only superficial, for all language rests upon a common "deep linguistic structure." So how could merely superficial aspects of a particular language ever give rise to deep differences in worldviews?

In his popular account of linguistics, *The Language Instinct*, Steven Pinker sets up a reductionist version of the Whorf–Sapir hypothesis and then proceeds to shoot it down. Pinker claims that Whorf–Sapir means that language is the *cause* of the way we think about the world. He then shows a variety of cases and experiments whereby human thought is able to get around the restrictions of language. But in fact the Whorf–Sapir hypothesis is not reductionist or mechanistic in the way Pinker makes out. It is really making a subtle point about language and worldview, but subtle points are difficult to test in the laboratory, and psychologists would prefer a simplistic version of the theory that can be verified or disproved.

There is great evidence to show that the way we perceive a given situation depends on the context in which it is presented—this includes everything right down to the rapid eye movements when we scan a scene. One aspect of this context is language, and thus language, perception, and worldview are inexorably tied together.

Human societies live in different ecologies and geographies. The natural world around us may dispose us to engage in hunting, farming, or fishing, into building villages or living a nomadic existence. These activities affect the way a society is structured; they determine who does what work and how kinship and ownership are structured.

It is inevitable that the particular languages and worldviews that evolve in different societies should go hand in hand, with each influencing the other, and here the word "influence" is very different from "cause"! It is often pointed out that, because Inuit hunt and travel in the far north, they have learned to discriminate between different properties of snow and have a correspondingly rich vocabulary. But this rich vocabulary does not "cause" them to see different aspects of snow in a mechanistic sense. It is just that they have a precise tool whereby they express their perception. Given such a tool they are more likely to make fine discriminations in this area of their experience.

We, too, in our technological world, have developed rich vocabularies. We have to deal with different sorts of machines, modes of conveyance, information systems, and so on and have the words to refer to them. Likewise doctors have a rich vocabulary of technical terms, as do lawyers and other professionals. Part of a doctor's training is to learn the names of all the bones in the foot. To know these names is to be able to discriminate between what, to the layperson, are just bones. Likewise lawyers use language to make fine distinctions over legal disputes. This does not mean that doctors or lawyers have different eyesight or logical abilities from the rest of us. Laypersons, too, can recognize bones and make logical arguments, but without the very fine tool of a technical language it becomes much more difficult to do so in any precise way.

Common sense tells us that flexibility, richness, and precision in a particular area of language enable us to convey discrimination with greater clarity. In turn, such discrimination allows us to see the world in a more clearly differentiated way. Jonathan Miller, in *The Body in Question*, points out that, to a layperson, the insides of the human body appear as a confusing mass of meat. However, doctors' training, which involves a combination of dissections and naming, teaches them to "see" the body in terms of various organs and their interconnections. As with the world's creation stories, the act of naming brings something into existence out of a background of chaos.

Of course a language is far more than a vocabulary. Blackfoot speakers emphasize the rich way they use verbs and discriminate between verb tenses. Living in a world of flux, their language is adapted to deal with constant transformation. By contrast, our Indo-European family of languages stresses the way nouns—objects in thought—are connected through verbs. These languages allow us to reify ideas and concepts and treat them as objects of thought. Bertrand Russell argued that a great deal of metaphysics comes about because of the way language is structured in that the subject of a sentence connects with the predicate. Because a predicate exists, and because "I" as subject relate to it, there is a persuasive tendency to treat the predicate as being in some sense real.

By contrast, the Blackfoot deal in process and when they need to refer to objects, they use verb forms. Names are more fluid, changing throughout a person's life. Blackfoot speakers have joked to me, "We're better adapted to dealing with quantum physics than you are!" This joke carries quite a layer of significance because it was exactly the same point made by David Bohm.

Bohm agreed with Bohr that we are inevitably suspended in language. He did not, however, agree with Bohr's conclusion that we are therefore forever blocked from discussing the quantum world. The problem arises, Bohm felt, because the quantum world deals in process, transformation, and flux, whereas European languages deal with the world in terms of nouns and concepts. What is needed is a true process-language, a language rich in verbs and in which nouns occupy a secondary, derivative place. Just as, in Bohm's view, an electron is a temporary structure constantly appearing and disappearing into the holomovement, so too nouns and concepts in Bohm's language are at best provisional and unfold out of verbs.

Bohm called this hypothetical language the "rheomode"—"rheo" referring to flowing. He even believed that it might be possible to use the rheomode in conversation and persuaded some students to try it out. The experiment was not a great success. Having been dependent on nouns all their lives the students began employing the rheomode's verbal structures to serve the function of nouns. It was only in the last years of his life that Bohm met with some Blackfoot people and realized that such a form of language had always been in existence.[5]

Language: Who Is Master?

In Lewis Carroll's *Alice in Wonderland,* Humpty Dumpty admonishes Alice when she asks about the meaning of a word. For Humpty a word means exactly what he wants it to mean, for, as he says, who is master—language or he? Bertrand Russell wanted to follow in Humpty's footsteps with his logical atomism, by bending language to his intentions and forcing it to mean exactly what he intended it to mean. As a young man Wittgenstein suggested that language creates pictures of reality and, provided we restrict our statements to those that are similar to scientific propositions, it may be possible to say something precise about the world.

---

[5]In fact Bohm's ideas on language were anticipated several decades earlier by the Argentine writer Jorge Luis Borges. In *Tlön, Uqbar, Orbis Tertius* of 1940, Borges writes of a country peopled by Idealists (in the sense of the philosophy of Bishop George Berkeley) who doubt the existence of objects and even the continuity in space and time of the experiencing subject. Those living in the south of the country employ a language consisting entirely of verbs.

Then, in later life, Wittgenstein realized that language truly has a life of its own. Poets are extremely careful about the way language is used and can spend days choosing the right word. Yet the very power of such poetry lies in the multiple resonances of words and the way they evoke a network of images, metaphors, and similes. It is these language games that present such a trap for philosophy, for the play of language creates confusion when philosophers begin to debate such issues as free will, consciousness, causality, and reality.

Language is a living thing. It allows us to play and be creative. It is well adapted to everything from the persuasive distortions of a politician or used car salesman to a teenager in love. Language is used for making puns and jokes, reciting epic poetry, composing a letter of condolence, or singing a folk song. Language is one of our finest tools, yet at the same time, to quote Bohr, we are always suspended in it so that we do not know which way is up and which is down. Our Western society is suspended in a language that favors nouns, while the Blackfoot flow along with a language rich in verbal forms. But Western science has now entered a new domain where noun-based languages may not be appropriate. On the other hand it is unlikely that we can transform our own spoken language to meet such a challenge. Thus was Bohr correct in arguing that we have reached a limit to knowing when we encounter the quantum world?

As this chapter has shown, linguistic certainty is another of those illusions of the early twentieth century that we have had to drop. Russell's "logical atoms" are incapable of coming together to create the richness of our world. We can never be totally unambiguous when we speak. We cannot pin down the world in words. But then language has so much more to offer us, and our lives are that much richer when language is not placed in a straitjacket. Maybe the next Niels Bohr will speak one of the Algonquian family's languages!

Five

THE END OF REPRESENTATION

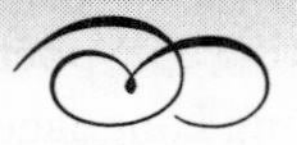

The previous chapter ended with some reflections on language and worldview. We will continue this general exploration by looking at the various ways we represent the world in everything from art and science to the way we speak. In fact, the way we picture the world within the mind deeply influences what we actually see and, in turn, how we think about ourselves and structure society. And by "seeing" I mean both vision through the eyes and vision within the mind, as in "picturing" the world mentally.

At first sight it seems rather extravagant to claim that the way we see the world influences what we think about ourselves. How can that be true? To understand this argument, let us begin with the Copernican revolution, which radically changed our sense of our position in the universe. Before Copernicus we located the earth firmly in the center of things. For more than 2,000 years human beings had pictured themselves as being contained, like a mandala, within a series of protecting spheres, planetary and divine.

The Christian vision, which dominated thinking throughout the Middle Ages, pictured humanity as the pinnacle of creation. Our task, according to Genesis, was to "subdue the earth." Christ's incarnation and crucifixion were not simply concerned with the fate of human beings but represented a cosmic event at the core of the universe. After the Fall, not only was the human race cast out of the Garden of Eden, but from that moment matter itself also fell from grace and awaited redemption. As Jakob Böhme wrote, "all of creation groans toward the day of fulfillment," and in Marlow's *Faustus* the doomed Faustus cries out, "See Christ's blood stream through the firmament." The entire cosmos circled around humanity. Human beings were the descendents of the Fall, and following that Fall the universe entered a state of expectant waiting.

All this changed with the Copernican revolution. Earth was demoted to become just another planet circling the sun, and humanity was removed from its throne at the heart and center of the cosmos. Later, following the invention of powerful telescopes, the sun was found to be just another star amongst countless billions. The Copernican revolution therefore produced a dramatic dislocation in our mental map of our place in the scheme of the universe. This shift in our picture of ourselves in relation to the cosmos gave rise to a fracture between inner, psychological space (where we felt ourselves to be) and the way we represented ourselves in relation to the new geometry of the cosmos.

In fact, this change in perspective had to be seen to be believed. As the art historian Martin Kemp points out, the Copernican revolution spawned a host of pictorial representations, from diagrams and drawings to mechanical models. To look at one of these diagrams or models was to perform an act of mental projection. It was as if we were now looking in from outside, as if we had abruptly shifted our position from living within the center of all that exists to watching the cosmos from its periphery.

A similar revolution in vision occurred when the first pictures of earth taken from space were published. Not only were they actual images of the earth, they also heralded a new and symbolic way of "seeing" our planet from within the imagination. That object, pictured as floating in space and seen from outside, was at the same time the home to all of us. Irrespective of our color or creed we were all indissolubly linked together within its global web of ecology.

One of my books, *The Blackwinged Night,* examined the dramatic changes in human consciousness that took place during the late Middle Ages and early Renaissance. It showed that a revolution in what it meant to be human within the world was precipitated by a radical change in the way people pictured and represented that world. The Middle Ages, for example, saw changes in representation in everything from perspective in painting to notation in music, from map-making to double-entry bookkeeping. Giving people new mental tools to represent aspects of the world around them meant that they could now externalize and objectify that world. Proceeding in this way they could treat the world as external to themselves and as something to be contemplated within the imagination. The world now became an object to be manipulated within the theater of the mind, rather than an external tangible reality. This also meant that people could gain increasing control over the world around them, yet always at the expense of a loss of direct involvement. The more we objectify the world, the more we are in danger of losing touch with that sense of immediacy felt by active participants in nature.

The Act of Seeing

From the moment we open our eyes in the morning, our acts of seeing are so automatic that we are barely aware of them. The world is simply "there." It is present to us. We see it without any apparent effort. Yet a closer examination of the mechanisms of human vision reveals to us that the act of seeing is highly intentional and by no means simply "photographic" as we may suppose. It is as if we are constantly reaching out into the world to caress its forms and textures with our eyes. What's more, a great deal of what we see arises out of what we expect to see. In other words, no real distinction can be made between seeing with the mind and seeing through the eyes; the two are inextricably intertwined.

Scientific studies tell us that the ability to see the world involves the integration of a variety of different strategies operating between the eye and brain. The most basic of these are "hardwired" within various areas of the brain. By the term "hardwired" I mean that all human beings, and many animals, have the same genetic instructions that allow for the development of similar sets of neural pathways responsible for vision. In other words, the first steps in seeing turn out to be the same for all humans. Signals from the optic nerve enter the brain, where they are routed to three separate centers: the midbrain, the cerebellum, and the visual cortex. The latter is itself divided into a variety of centers; in each of which vision is doing something quite different. Midbrain vision, for example, is also present in much simpler organisms such as the frog. At this level it is probably true to say that we don't actually "see" anything, in the sense of registering a visual scene in our conscious awareness. The more primitive functions of the midbrain are instinctual. When a fly enters a frog's visual field, the frog's tongue darts out as a reflex action. In human terms the frog could not really be said to have "seen" the fly just before shooting out its tongue.

Vision, in the sense of actually seeing things, begins with processes in the various areas of the visual cortex. One of these processes involves seeking out edges. Discerning the edge of an object is important in trying to determine its outline. Other strategies are used to pick out moving bars, fields of color, areas of movement, and so on. (When researchers design robots to recognize objects they exploit similar strategies in their computer programs.)

These first stages of vision therefore involve the simultaneous, but separate, processing of information received from the eye. At this point the brain does not yet "see," for example, a red and yellow box falling from a window, but rather a series of edges, a field of movement, some areas of color, and so on. Next the various outputs of visual material are integrated to form a visual whole. It is only at this point that we "see" a falling red and yellow box, or a blue car driving away from us, or a man waving.

The description above approached vision from one side only, for seeing is also very much about the act of doubting. As the brain attempts to integrate the visual clues it has collected, it rapidly makes a series of guesses: Is that a blue moving car or simply the wind disturbing the reflection of the blue sky on a lake? Is that dark patch a shadow or could it be the black fur of a stalking animal? Is someone hiding in that bush or is it merely a pattern of foliage?

The eyes are constantly looking for certainty while the brain deals in doubt. This is where the intentionality of vision comes in. From moment to moment the brain seeks out what is relevant from within the information that it is receiving. It is constantly making hypotheses about the world around it and, like a Popperian philosopher, it needs additional information in order to reject some of these hypotheses while provisionally accepting others. For this to happen the brain must instruct the eye where to look and what to look for. Therefore signals are constantly being sent to the muscles around the eye in order to direct it to explore various areas of the visual scene and gather more information.

Not only is information constantly streaming up the optic nerve toward the brain, but equally important, a series of questions and interrogations is flowing downward from the brain toward the eye. These two streams meet at points all along the optic nerve. The signals coming down from the brain interrogate upcoming raw data and ensure that only significant information reaches the visual cortex—in other words, those answers that help the brain reject certain visual hypotheses and provisionally confirm others.

An analogy would be a hypothetical filtering system attached to your email program. It would scan the content of each message at the moment it is being downloaded from your server. Messages from close friends and from colleagues relevant to your work get through and are displayed on your screen; but junk email, advertising, chain letters, and so on are dumped in the Trash bucket. Moreover, the system is an intelligent one, for it is constantly scanning the work you have been doing on the computer so as to detect what will be significant to you from moment to moment. In this way you only receive messages of immediate importance and need not bother about the rest.

It is the same with vision. The brain doesn't want to be overloaded with everything the eye is detecting. It is only interested in information relevant to the scene it is attempting to build up, as well as monitoring this scene in case anything of significance changes. This visual data that finally reaches the brain helps to create a hypothesis about the world outside. In turn, the brain now directs the eye to move and collect new data that will help to confirm that hypothesis or resolve visual ambiguities.

Seeing within the Mind

Vision therefore involves a constant movement between the generation and resolution of doubt. But this means that a great deal of what we "see" must already be present in the brain in the form of assumptions based on what we have already learned about the world and the way it works. Indeed, what we see is not so much what lies in front of us but what has been created out of memory and the visual strategies of the brain. If we begin to make out a person's face against a background then we immediately expect to see two eyes, a nose, and a mouth. If the person is wearing a mask we receive a visual shock indicating that something is badly wrong. As we walk out the door in the morning we unconsciously notice the position of the sun in the sky, and our brain is alerted to pick out shadows falling in particular directions and to distinguish them from oil stains on the road or patches of dark soil. In short, a large part of what we see is what we expect to see.

This explains why we "see" faces and figures in a flickering campfire, or in moving clouds. This is why Leonardo da Vinci advised artists to discover their motifs by staring at patches on a blank wall. A fire provides a constant flickering change in visual information that never integrates into anything solid and thereby allows the brain to engage in a play of hypotheses. Conversely, the wall does not present us with very much in the way of visual clues, and so the brain begins to make more and more hypotheses and desperately searches for confirmation. A crack in the wall looks a little like the profile of a nose and suddenly a whole face appears, or a leaping horse, or a dancing figure. In cases like these the brain's visual strategies are projecting images from within the mind out onto the world. We can also observe some of the strategies of the visual system at work when we are in a high fever. During delirium, regions of the room may seem to move, the ceiling may appear to fall toward us, or figures may leap out of walls.

While some of these strategies are hardwired, or genetically determined, much depends on how we grew up and the environment that surrounds us. The city dweller learns to see a very different world than does a desert nomad or an Inuit in the high arctic. Each would feel lost in the other's environment. Seeing involves intentionality through constant acts of doubting the world and then looking for ways to resolve those doubts. In turn, the visual hypotheses we make about the world have a great deal to do with the context in which we are placed from moment to moment, a context that involves not only the particular location in which we find ourselves but also the whole of our society, right down to the language we speak.

In one experiment, photographs were presented to a variety of subjects while the experimenter introduced a topic of conversation. Depending on the context of the conversation the subjects "read" the same faces in radically different ways, discovering in them everything from benevolence to criminality.

One of the most striking examples of the way we construct meaning out of contexts is the "Kuleshov effect" discovered in the early days of the cinema. The Russian director and teacher Lev Kuleshov filmed the neutral expression of a well-known actor and then edited it into a series of shots including a bowl of soup, a dead person, and a child playing. When audiences were shown these edited sequences they "read into" the actor's expression the feelings of hunger, sadness, and affection. In fact those who saw the film praised the subject for his acting ability.

The art critic John Berger demonstrated a similar effect during his television series *Ways of Seeing*. Berger photographed portions of Caravaggio's painting *The Meal at Emmaus* (showing Christ and his disciples) and edited them together. He then used this edited sequence in his television documentary to the accompaniment of background music. In the first case the music was from Bach's *St. Matthew Passion*. In the second case the identical sequence was accompanied by comic Italian opera. In one context viewers "saw" a deeply religious painting, in the second, a group of Italians enjoying a meal.

Language and Vision

Contexts influence how and what we see, and, as we saw in the previous chapter, language is a particularly significant context and is therefore deeply connected to the way we see the world. As we have already noted, part of a doctor's training involves memorizing the names of body parts; when the knowledge is combined with dissections and anatomy lessons the medical student gradually learns to "see" the various organs and components of the body. The inside of the human body would look like a messy collection of meat to an untrained observer. But having been trained in naming, a doctor sees something quite different—an interconnection of organs, blood vessels, nerves, muscles, and so on.

Likewise, we see patterns in the night sky because we have been told stories of the constellations and have learned their names. Other cultures use different names and stories and so see diverse patterns.

As we saw in the previous chapter, the Blackfoot language is very much "verb based," in the sense that verbs form the most important part of the language, with many nouns being secondary and derived from verbs, and so it is not surprising that the Blackfoot live in a world of constant flux and transformation.

It is not so much that particular languages evolve and then cause us to see the world in a given way, but that language and worldview develop side by side to the point where language becomes so ingrained that it constantly supports a specific way of seeing and structuring the world. In the end it becomes difficult to see the world in any other light.

Creativity and Doubt

The ways we represent the world, in everything from language to art and science, deeply influence the ways we structure our world and understand ourselves. During the twentieth century many of these means of representation underwent a change from certainty to uncertainty, and today our world is more tentative and open to doubt and uncertainty. This lack of fixed strategies means that there are more ways to explore the world and that we must therefore exercise a deeper sense of the responsibility that goes along with this freedom.

This lack of certainty may be one of the reasons why ours is not an age of great art and literature. There are no all-encompassing statements to make or great contemporary myths to relate. Our world lacks the sense of confidence and certainty necessary for a Bach or a Michelangelo. In a period of transition, when everything is open to question, our greatest creativity may lie not so much in producing works of art as in building new social structures and more stable and sustainable relationships to the natural world. It is only after this period has passed, a period that may last well into the twenty-first century, that a new context will be created, one in which new myths and new artistic endeavors are possible.

Painting

Changes in the way we see the world are also evidence of changes in human consciousness. This is most easily seen by looking at paintings, particularly those of different cultures or those made centuries ago. They are an important way of discovering how different people structure their world. Today we have the additional benefits of photography, film, and television. After all, who under the age of 50 can imagine what it was like to live in a world of black and white pictures and movie newsreels? Looking at films made in the 1930s, with their home interiors, clothing, big cars, soda fountains, and small-town life, we see a profoundly different world. It is a world that seems grainier, more starkly etched, more direct and simple. By contrast, in contemporary cinema it is increasingly difficult to detect what is real and what has been constructed by computers and postproduction processing. Where once "seeing was believing" today we can no longer be sure of the actuality of a TV news clip or newspaper photograph.[1]

---

[1]In fact, in some of the great classical photographs of the past all is not what it seems. Robert Doisneau's famous image of two lovers kissing was posed, as was that of a young woman walking past wolf-whistling Italians. St. Mark's Campanile, appar-

In George Orwell's *1984*, newspapers of the past were constantly being written in accordance with the current proclamations of Big Brother. If party members were disgraced, their names disappeared out of newspapers and the record books. If Big Brother claimed that steel production had increased (when of course no such thing had happened) back numbers of newspapers were rewritten to show much lower production figures in earlier years. *1984* was a work of fiction, but have you noticed that many of the great figures of the past are no longer smokers? Once the cigarette and the haze of smoke it created were romantic images for an actor, writer, or musician. Today smoking is downplayed, and many of those old photographs have now been carefully treated to remove the cigarette! The U.S. Postal Service, for example, removed the cigarette from the photograph of Jackson Pollock used for its 33-cent stamp. In this way, writers, composers, and film stars have become, retroactively, nonsmokers. How much more of the past will be re-imaged through the eyes of the present?[2]

Photography and cinema are relatively new inventions, so if we want to take a broader focus on the way societies have represented their world it must be through painting. Paintings on cave walls and pottery are among the earliest records of human existence. Paintings can be found in nearly all of the world's civilizations—on plastered walls, the desert sand, wood panels, sarcophagi, furniture, papyrus, parchment, paper, and, eventually, canvas. Looking at paintings from different parts of the world and from different historical periods tells us something about the world in which people lived. Paintings from Tibet and India speak to a world of powers and energies. They are not so much depictions of particular gods as diagrams of spiritual energies

---

ently shot at the precise moment the tower collapsed into rubble in 1902, and Yves Klein's "leap into the void" from a Paris window, were both created out of a montage of photographic images.

[2]Some years ago the magazine *BBC Music* had a cover of Leonard Bernstein at the piano with a cigarette in his mouth. The accompanying editorial explained that for a time they had considered removing the cigarette and its wreath of smoke but, in the end, felt that heavy smoking was so much a part of Bernstein's image that, despite "political correctness," it just had to stay in.

and doorways allowing people to enter into other worlds. West Coast totem poles speak of a family's relationship to creators and keepers of the animals. And, as the eye moves back and forth across a Chinese scroll, we sense a different way of structuring space and time.

Paintings can employ many different devices to achieve their ends. Some are emblematic and use colors in a heraldic way to represent events and historical figures. Some tell the story of the life of a pharaoh or king. Some are highly diagrammatic and represent the essence of objects rather than their external visual appearance. Some paintings are devices used to instruct or to relate myths and legends, a kind of early comic book illustration. The paintings of Australian aborigines are journeys, maps of the land and locations of power where the ancestors walk during dreamtime. In other cases, painting is an expression of pure joy, as when a variety of motifs are used to brighten up furniture, caravans, barges, and so on.

In each instance, line, color, and form have been used to create diagrams, symbols, maps, and illustrations of particular aspects of the world. Paintings indicate the ways particular peoples have chosen to structure their world and, in turn, paintings are the representations of this structuring. They could be thought of as providing ways of seeing into the past, into the landscape, into the energies of nature, and into the powers of the gods, or simply an expression of delight at the visual appearance of the natural world.

The vast majority of the world's paintings employ one or more of these devices. This is why, from the Renaissance to the end of the nineteenth century, Western high art is particularly unique. For several centuries, and to a greater or lesser extent, it downplayed these other approaches to painting because of its concern to represent the world in a highly illusionistic way. Its aim was to perfect ways of representing the appearance of things at single instants of time.

This is not to say that other civilizations did not engage in illusionistic representation. After all, portraits painted on the lids of Greek and Roman coffins are quite realistic. Nevertheless, since the Renaissance, Western art has been concerned with illusionistic representation. In part, I believe, this stems from an increasing sense of having objectified the world within the mind. Rather than living inside of nature, human consciousness had begun to abstract itself from the world. It pictured itself as standing outside and looking in. Rather than being an active participator within the cosmos it had become an observer and contemplator. Naturally this required an art that would complement this new attitude of mind and a means of representation adapted to an objective gaze.

Contemporary with this change of seeing came the invention, in Renaissance Florence, of geometrical perspective. There had always been a number of ways of expressing the solidity of things and their position in space. For example, making objects larger or smaller, with the closer object overlapping the ones behind. Other clues as to distance come from the haze, the blue cast of distant mountains, the changing sizes of tiles or patterns on a carpet and so on. Point perspective was dramatically different and led to a particularly vivid form of illusion.

Before perspective came on the scene the Sienese painter Duccio could paint the *Madonna Enthroned* as if seen from several different viewpoints. Part of her throne is seen from the left, another part from the right. Duccio was giving us more of an all-round view of the throne by integrating several glances into a coherent whole. Western painting would have to wait for Picasso and Braque before this device would be used again in any coherent way.

By contrast, perspective employs only a single viewpoint. It is as if the world is being seen through a window, the edge of the canvas or panel being the window frame. In his frescoes for the Scrovegni chapel in Padua, Giotto, one of the first painters to experiment with perspective, made explicit reference to this device. On the right and left of the altar he painted archways through which could be "seen" the transept of the church with a hanging lamp. To complete this illusion we "see" a different angle of this transept from the right- and left-hand viewpoints. It is as if Giotto had punched a hole through the wall and we are seeing a real extension to the chapel.

Of course not every painter slavishly adopted the perspectival grid or preserved the illusion of the world seen through a window frame. Caravaggio knew the rules of perspective well enough to subvert the whole illusion in particularly dramatic ways. Rather than framing his figures and objects, he has them appear to leave the picture plane and leap out into the viewer's space. With Caravaggio, bowls of fruit are about to fall, chairs crash down, arms are flung into our faces. Nevertheless, all this was done to heighten drama by representing the visual actuality of things in naturalistic ways.

Perspective is a marvelous tool for producing a particular type of illusion of reality. Yet one should never forget that, even in the most naturalistic of paintings, it remains an optical trick. It does not really represent the way we see the world but, because of the prevalence of perspectival paintings over the centuries, we have come to assume that they are indeed the only realistic way to depict the external world. In fact, it depicts the world as seen by a one-eyed person with her head clamped into position and at a single instant of time. It is the same system of representation used in the camera, where a lens focuses an image on a photographic plate and the shutter is opened for a fraction of a second. Yet, as we saw at the start of this chapter, the human eye is never fixed. It is constantly scanning the visual scene. Likewise the head moves to take in different glances while the brain integrates all this information into a coherent whole. When we shift from thinking about realism as an objective way of representing the external world, to asking how we can convey our subjective *experience* of seeing the world, then many other ways of painting open up for us.

The painter David Hockney argues that the device of perspective originated out of artists' desire to paint the crucifixion. Of all forms, a crucifixion portrays a precise instant of time, in many cases the moment when Christ gives up the ghost. Likewise, it is natural for Christ to be at the center of the painting, flanked by the two thieves and with Mary and the disciples looking up at him. Each gaze is focused on this central figure; time is frozen and perspective captures this scene most perfectly.

Whatever its origins, perspective continued to dominate Western painting through the centuries. It was applied not only to religious but also to secular subjects. Perspective also provided a realistic way of gathering a number of people together in a group or crowd. Rembrandt's *Night Watch,* for instance, provided a way to satisfy the egos of rich burghers in a group portrait.

In the hands of Dutch and Spanish painters the illusionistic portrayals of fruit, vegetables, and tableware gloried in the richness of textures and surfaces and hinted at a spiritual essence within the natural world. Landscape painting could be used as a delight to the eye, an experiment in observation, or an expression of a landowner's wealth.

During the nineteenth century many paintings were given over to storytelling. Victorian narrative painting often presented a moral tale or expressed some socially suitable sentiment. Clearly this was most effective when the illusion of reality was preserved. Again the means of representation perfectly complemented the mindset of a society. Victorian England was hierarchical: individuals had their place, and law and government were to be respected and heroes elevated. The poor and those displaced from the countryside were an inevitable consequence of industrial expansion and so deserving of charity. Such figures were portrayed sentimentally in painting and story. Heroes, on the other hand, demanded vast canvases brimming with figures, such as Benjamin West's *The Death of Wolfe.* Living figures could be elevated by painting them wrapped in cloaks or Roman togas. The pre-Raphaelites, rejecting materialism and the ugliness of industrial Britain, portrayed a return to some sort of Eden by means of highly detailed and realistic paintings.

In France the Revolution and the subsequent rise of Napoleon represented a serious disruption of the prevailing social order. At such times (as in Nazi Germany and Stalinist Russia) artists were expected to be sober and, in their canvases, to give authority to a regime through reference to well-established classical models. France found its court painter in Jacques-Louis David, who gave them allegories of contemporary events as if they were taking place in Rome or ancient Greece. In this case it was not so much an artist finding a visual means of representing the worldview of a society, but rather of supporting a fiction or fantasy of what that society would like to be. David's approach worked because he was a painter of genius. Hitler and Stalin were not served so well by their official painters. The supposed "heroic" portraits of the German dictator, for example, look to us today like a ridiculous little man dressing up in an attempt to make himself larger than life.

One of David's masterpieces, *The Oath of the Horatii,* was completed just before the Revolution. It depicts the moment when the aged Horatius hands his swords to his three sons who swear to defend Rome against the city of Alba. To the right of the painting the women of the family are weeping, one is engaged to an Alban, and the other, herself an Alban, is married to one of Horatius's sons. This painting became a rallying sign for revolutionaries. They saw it as a call to take up arms and fight for the new order, even if this meant division between family and friends.

The extent to which neoclassical representations can serve a society as propaganda for its worldview can be seen by the scandal unleashed by a painting that did the direct opposite. Theodore Gericault painted his *Raft of the Medusa* in 1882, following the restoration of the Bourbon monarchy.

In 1816 the French frigate *Medusa,* bound for Senegal, drifted off course while the crew were celebrating the crossing of the equator. The ship struck a reef at high tide and began to sink. Since there were insufficient boats for the entire crew to escape, a raft was made with the intention of towing them to land. This raft was supposed to serve for the 152 crewmembers but as soon as they boarded, it began to sink so that food and provisions had to be cast overboard. Naval officers refused to take command of the raft and, after towing it a short distance from the reef, it was cut adrift with the crew still on board, the captain and officers escaping in the ship's boats. The raft drifted for 12 days, and, without food, the desperate men were forced to resort to cannibalism in order to survive. Of the 152 crew abandoned by their officers, only 12 survived to tell the tale. Gericault paints the moment when a ship (a rescue ship, perhaps?) is sighted on the horizon as one of the crew waves his shirt to attract the ship's attention.

The painting precipitated a national scandal and exposed corruption at many levels of the government. Not only had officers abandoned their men, but it turned out that the captain, an old and incompetent man, had obtained his appointment through political connections.

Whether a painting celebrates a society's image of itself or exposes the inherent falseness of that image, the fact remains that all these paintings, based upon a perspectival geometry, are about a form of certainty. They say, "this is the way the world is," or "this is the way the world should be." There is no room for doubt in such paintings, no place for paradox or complementarity. Such paintings seek to represent reality, yet at the same time they really do not engage the essential way in which we actually *see* the world.

Art as a Scientific Theory

Much of the world's nonrepresentational art is concerned with visual designs, symbols, signs, maps, diagrams, indications, records, and calligraphies that delight the senses and stimulate the eye. At the same time they are often pointing to something that lies beyond them. They may express beauty, pattern, harmony, and order but also a sense of the sacred and the numinous and a connection with all living things and the energies of heaven and earth. Rather than being purely concerned with the visual surface of things, they point toward their inner structure, to an underlying order of the world, a reality beyond appearances. Islamic art, for example, employs highly repetitive patterns in its tiles and ironwork. The meaning of these patterns is that they lead toward the infinite—not so much an infinite that lies away from us "out there" but the infinite within, the infinite of the endlessly divisible and repetitive. Islamic art is a device for the mind's eye. It is a tool for transporting human consciousness toward pure contemplation of the boundless infinite. Similarly, the mandalas in Tibetan art are a way of bringing consciousness to a still center and placing the mind within its proper relationship to the powers of the cosmos.

In this sense such art has something in common with a scientific theory. A theory is not so directly concerned with reality but rather with a *model* of reality. In turn, a model is not the thing in itself, for it always points beyond itself. Think of a toy train. A model train runs on tiny railway tracks. It has a smokestack, a tiny driver, and all the appearance of a real train. It looks like a train and, in its motion, evokes a train, yet at the same time it is *not* a train. There are no living passengers in the carriages and no real fire in the firebox. A toy train incorporates some of the elements of a real train while neglecting others.

Likewise a scientific theory is a model of the real world, a model in which, for example, there is no friction, no air resistance, a model in which all surfaces are perfectly smooth and all motion is totally uniform. It refers to a world in which everything has been idealized. There is an old joke among physics teachers that to solve a particular problem you must "take an elephant of negligible mass." To a layperson this is an absurdity but this is exactly the sort of approximation and simplification that is sometimes made in order to apply scientific theories.

Of course corrections and additions can be made to any theory to take into account, for example, the elephant's mass. But elegant theories, like beautifully built model trains and airplanes, are of necessity simple. They say, "I am not the thing in itself but I point toward that thing." Likewise a Hindu or Tibetan painting says, "I am not a god. I am not even a representation of a god (in the sense of a photograph of a person). Rather I point toward something that lies beyond myself, and that which I point to may lead you to an experience of the god."

Such art always preserves the tension between what is and what is not. Great art, such as that displayed in a Russian icon, is a frame or container that holds this tension between two worlds and, in so doing, becomes charged with numinous power. By contrast, representational art does not hold such a duality. It doesn't say, "I am not the Death of Nelson but point toward an important historical event." Rather it says, "I'm just like the Death of Nelson. If you had stood in that particular spot and at that particular moment in history this is more or less what you would have seen." It says, "If you look at me you will give your eye the actual experience of the surface and the texture of a bowl of fruit. I am the exact image of the appearance of things in the world."

Yet in the end a painting remains a painting. Naturalistic and illusionistic painting calls on us to enter into Coleridge's "willing suspension of disbelief." Paintings ask us to collude with them and imagine that we are looking at a scene through a window or standing on the deck of Nelson's *H.M.S Victory*. Indeed, in the nineteenth century some painters produced vast panoramas that stretched around an entire room so one was enveloped in the illusion of a great outdoor scene.

For such an illusion to work, visual doubt must be abandoned, or at least bracketed out of the equation. In the end, all this was to change as society changed.

Impressionism

By the middle of the nineteenth century a new social class had emerged in France. These were neither rich nor poor but the petite bourgeoisie who owned shops, worked in factories, lived in the new suburbs, and delighted in the dancing at Bals Musette or picnicking by the Seine. Theirs was not the world of the official Salon with its vast historical canvases. It was a smaller world, more immediate and concerned with everyday objects. Soon it had a group of painters who would complement their more modest and less blatantly "heroic" worldview.

The Impressionists painted out of doors, or at least began many of their paintings in the open air. This was made possible by the invention of paint in tubes, an example of that constant link between art, science, and technology. Impressionist canvases are generally small enough to be carried outdoors, along with an easel, and the motifs painted are those of everyday life—riverbank strolling, picnics, dances, cafés, and a train station.

It is a great oversimplification to see Impressionism as a single consistent movement in art. The fact is that for a few years a small group of painters were sufficiently like-minded to exhibit together and sometimes even paint side by side. But fairly soon each of them was to pursue his or her own particular vision and approach. For this reason it is easy to identify the work of individual painters compared with, for example, trying to figure out if a cubist painting is by Braque or Picasso. Yet there is one thing that all held in common, and that was the significance and immediacy of visual sensation.

The greatest consistency to this original vision of Impressionism lies with Claude Monet. He was concerned with the immediacy of things and the act of seeing. The validity of the visual impression came first, rather than attempting to fit what had been seen into some predetermined structure—a classical arrangement of figures, for example, or strict geometrical perspective. Monet and the others wanted to paint what they actually saw, rather than what they expected to see, or hoped to see.[3] Cézanne said of Monet that he "is only an eye but, my God, what an eye!"

To explore the effect of changing light, Monet painted the same object at different times of the day. Rather than seeing it as, for example, the same cathedral face but lit in different ways, Monet was actually seeing an entirely different scene each time he painted. Light and color were essential aspects of what lay before him. Rather than color being the surface attribute added on to an object, or light being the means by which that object was seen, light and color were coexistent entities of equal importance as matter itself. They could never be divorced from the object—the tree, the cathedral, and the locomotive—but were an inseparable part of participatory seeing.

When the Pre-Raphaelites painted their highly detailed and hyperrealistic paintings they would stand close to each object to observe its actual color. By contrast, Monet was aware that as he looked away from one colored object, a fugitive sensation would impose itself on objects nearby. (Stare at a red object then look away to a blank white wall and there will be a green afterimage.) Likewise shadows were never black but made of a complementary color. Rather than attempting to bracket out these various effects as being accidental and unimportant when compared to the surface appearance of things, Monet felt they were all equally worthy of his attention. He even went so far with these "fugitive sensations" as to include the "floaters"—tiny bits of fat that move within the eye and cross our field of vision—in his paintings. At the height of his powers Monet worked like a scientist, constantly observing, experimenting, and seeking to set down the truth.

Cézanne was equally forthright when it came to the search for visual truth. Cézanne's aim was to move beyond Impressionism as a means of setting down visual sensations, by combining it with a new

[3]The most common mistake among those who enter art school is that they paint and draw what they *think* lies before them, rather than paying close attention to what it is they are actually seeing. A traditional training is about opening the eyes and learning how to see.

order in painting. He spent the rest of his life searching for this order. Again and again he remained unsatisfied with what he had painted and only a few of his works are "achieved" in the sense that he was willing to add his signature to them. As he painted, Cézanne would move his head, interrogating the scene and seeking to resolve ambiguities in what he was seeing. In many cases a visual doubt remained, and, being an honest man, Cézanne allowed these visual doubts to remain on the canvas rather than "correcting" them or attempting to resolve his perceptual uncertainties and ambiguities.

Cézanne wished to remain true to his "little sensations," so rather than painting over what could be taken as a "mistake," or a trial attempt at depiction, he let the mark stand and added another nearby. And so we notice the tentative nature whereby he paints the branch of a tree, recording the various sensations of where that branch could be located. He doubted the nature of a piece of foliage and told us that this visual sensation could mean a group of leaves in the immediate foreground or a bush in the background. The ambiguity remains on the canvas—two possible interpretations, complementary visions, doubt as to the nature of visual reality.

And so Cézanne returns us directly to the act of seeing within the eye and mind, to the constant process of doubting, making hypotheses, again doubting their validity, rejecting some and provisionally accepting others. Picasso and Braque, who came after Cézanne, experimented for a time with versions of cubism whereby the different possible impressions we receive as we walk around an object are recorded and integrated into a flat, two-dimensional whole. It is probably no coincidence that Monet's wish to return to direct visual experience, Cézanne's doubt as to what his sensations were telling him, and Cubism's attempt to integrate different possible viewpoints in time should coincide with a general change of Western consciousness whereby, as we have seen, doubt, relativism, and a lack of certainty entered in many different ways.

I have referred before to the idea of a change of consciousness during the twentieth century. I have also drawn attention to the many different ways in which doubt entered physics, mathematics, philosophy of language, and now art. There are many other examples of movements in one area being reflected or paralleled in another. Toward the end of the nineteenth century, Georges-Pierre Seurat transformed painting into a series of dots of pure color, almost as if in anticipation of the way Max Planck, in 1900, would transform light into individual quanta. Cubism reintroduced time into the space of the canvas just as Einstein and Mach reintegrated time and space. Two centuries earlier, while Dutch painters were exploring the way light enters a room, Newton was allowing a crack of light to enter his study so that he could break it down with a prism. There are many more parallels that could be mentioned in which ideas seem to complement each other and appear at the same time in many different fields.

In some cases there may exist a causal relationship between one area and the next. When computer engineers started to display fractal forms, artists in turn began to study inner complexity and use fractal forms in their work. When the chemist Michel-Eugène Chevreul made an analysis of the psychology of color, Seurat immediately made use of these discoveries in his paintings. Yet, in many other cases, the artistic innovation came first, or there was no direct or traceable link between the innovations in art, science, and literature.

How could this be? Why should so many remarkable parallels exist? I have thought about this for much of my life and sought a variety of avenues that will "explain away" such coincidences. Finally I have been driven to conclude that they are the manifestations of an actual change in human consciousness involving, for example, a change in the way we "see" the world. At a certain point "the time was right" and "something was in the air." Human consciousness was at a critical point and this potentiality for change was first picked up, symbolized, and expressed by writers, artists, and scientists in their respective fields. Rather than the one influencing the other directly each was picking up and manifesting the seeds of change.

Maybe these parallel manifestations in art, science, literature, and other fields should be more properly called "synchronicities." As Jung defined it, a synchronicity is "the coincidence in time and space of two or more causally unrelated events which have the same or similar meaning" or more simply "acausal parallelisms." In the popular imagination synchronicities are associated with remarkable coincidences that happen to people. One dreams of an old school friend and on the following day receives a letter after 20 years of silence. But synchronicities can occur at the collective level of society itself.

Our rational, waking, reasoning life is only a small part of our total experience. Likewise, individual, personal experience is only a small part of what is available to us. At the collective level we all dip into what has been called the zeitgeist of the times. We have access to a dimly felt edge of where consciousness is moving. Some are more able to plug into it than others. As Stravinsky once said, "[T]he artist isn't ahead of his time, the public are behind theirs."[4]

Postmodern Values

In a world in which absolute certainty has been left behind, art itself became many different things. For a time it viewed the canvas as an object in its own right. A painting no longer referred to something lying outside itself, but rather, for a Jackson Pollock, for example, the canvas itself became the arena of action. The canvas did not stand for something beyond itself; the canvas had become the thing in itself.

Art was also performance, it was social action, it was a breaking of boundaries between the picture, the frame, and the wall, between the gallery and the world outside, between the artist's products and the artist's life, between what it meant to be an artist and what it meant to be a citizen.

Art broke into a multiplicity of activities and products. More and more it encouraged each one of us to be makers of art ourselves. As the sculptor Anish Kapoor asks, "[D]oes the work of art lie in the stone, in the mind of the artist, in the eye of the viewer or in the space between?" It is in the participatory act of seeing that art is born. As the German artist Joseph Beuys put it, "We are all artists." In this sense we are all helping to create the world.

---

[4]I'm a little reluctant to make such critical distinctions between public and artist, but it is certainly true that some people come into life with highly tuned sensibilities and act as beacons for the rest of us.

Yet this latter statement brings us to the heart of the postmodern dilemma. If there is no certainty anymore; if everyone is an artist and if art is a multiplicity of activities, from painting to performance, from texts on a wall to a walk across a field, from decorating a telephone box to living out other people's fantasies on the Internet, does this mean that everything has collapsed into a formless relativism? Are there no longer any values, judgments, or standards in art?

Of all the things around us, from crime in the streets or pollution of our rivers, what most upsets the ordinary citizen is a row of bricks in an art gallery or an enormous canvas apparently painted in a single color. "Because I could have done that just as well, how can that be art?" is the general reaction, and "if *I* did that no one would pay me a million dollars." The result is that the contemporary art industry looks like an elaborate con game set up by artists, dealers, and gallery curators.

To a certain extent this criticism is correct. There is a great deal of confidence trickery in the art market and, as any con artist will tell you, it is the greedy and acquisitive who are the most readily fooled. But the fact that some people collude does not mean that contemporary art itself lacks any value. Of course it has values. The issue for many laypeople is, Who sets these values? In the past the public could look to the salon, the national galleries, and the art experts to be told what was good and what was bad. But what use are art critics today? How are ordinary people supposed to find their way through all that fog of convoluted jargon that is being written about art?

Art today is particularly diverse. No single authority is willing to tell the layperson "this is good art and that is bad art." But this does not mean that all judgment is for naught. More and more the onus is being put on the viewers, or participants, to respond to art, to make their own judgments and break down the barrier between art and artists on the one hand, and themselves on the other. The viewer–participants have a right to question and to refuse to accept what is put before them. But to exercise this right they must at least be willing to meet the artist part way and to assume some of the responsibility of what is being shown, performed, or said.

And so the postmodern dilemma has two sides and two faces. It is telling us that times have changed and we can no longer go to an art gallery or exhibition with that comfortable sense that we are "being educated," "made better," or "given a dose of culture." It is no longer sufficient to wander from room to room, reading the names of the artists on the little plaques beside the paintings and only stopping in front of work if it is by a "famous artist." Engaging contemporary art means engaging our own doubts. It means no longer taking things for granted and at face value. It means being open to new experiences and accepting discomfort rather than always expecting to see or experience what is familiar and easy. Art has opened us to doubt, and along with this doubt comes a great deal of responsibility.

The World as Surface

Finally I want to return to immediate visual experience and the ways in which one branch of art continues to challenge us with the surface of our world. By this I mean photography. A few years ago I was invited to write an essay on photography and science for the Museum of Contemporary Photography at Columbia College, Chicago. The manager of the collection, AnJannette Bush, kindly sent me a large package containing slides from much of their collection representing photography over the last 50 years. Most of what I looked at was not scientific photography as such, but representative of the work of documentary, commercial, and "art" photography over the past few decades. What struck me as I looked through the collection was a common denominator in the way very different photographers were seeing the world. It seemed to me that our world, or at least our immediate city environment, has become one of reflections and surfaces. It is a world in which we can never be too sure of the tangibility of things. We walk past an office building and wonder if that is a tree growing inside or if it is a reflection of the outer world in the building's glass windows. Everywhere we look we see reflections and superpositions. Glass, plastics, computer screens, virtual reality, transparencies, television advertising, and packaging all present us with intangible images and transitory objects. Advertising is composed of overlays, montages, and ambiguity.

In many ways these mutual reflections are themselves representations of changes within the fabric of modern society. Philosophers such as Jean Baudrillard argue that the twentieth century's obsession with material consumption led to an increasing replacement of real objects with signs and images. Will this trend continue? or will a future society reconnect with the real? How far can we tolerate our uncertainty? For how long will we continue to accept the consumer image in place of the real? In moving from certainty to uncertainty, how will we begin to represent and envision our new world?

Six

FROM CLOCKWORK TO CHAOS

We have already seen a number of ways in which that pivotal year 1900 stood as the watershed between certainty and uncertainty. This chapter introduces yet another of these revolutions—the introduction of chaos into the heart of science. Today chaos theory, along with its associated notions of fractals, strange attractors, and self-organizing systems, has been applied to everything from sociology to psychology, from business consulting to the neurosciences. As a metaphor it has found its way into contemporary novels. As a technique it is responsible for the special effects of so many movies.

Chaos theory has become ubiquitous, but to discover its origins we must go back to 1900 and a study made by the mathematician and philosopher Henri Poincaré. Poincaré was investigating another of those certainties, one that the human race had lived with since the beginning of time: "The sun will always rise in the morning and set in the evening." In questioning the inevitability of things he was challenging our certainty that the earth's orbit around the sun will continue to repeat itself. In his research Poincaré was touching something very deep, no less than civilizations' entire way of understanding time and what it means to live within a cyclical nature. In doing so he was touching the seeds of chaos, and maybe this is the reason that the term "chaos" and the notion of a chaos theory has proved to be so disturbing to a mind that seeks order, regularity, and predictability.

The Womb of Time

Early human societies were embraced within the rhythms of nature. They lived with the rising and setting of the sun, the heat of midday and the cool of the evening breeze, the long days of summer and frosty nights of midwinter. Nature's rhythms were so ubiquitous that humans bent to their demands.

Then at the end of the thirteenth century the first mechanical clocks appeared on public buildings and, in towns at least, people became aware of a new quality of time. It was a time measured mechanically, a time divided and subdivided into equal proportions. No longer did it matter if it was winter or summer, if there was plowing or harvesting of grain to be done, for the mechanical hours each lasted the same duration. Irrespective of work to be completed, or the amount of daylight remaining, clocks ticked away the hours and minutes equally throughout the seasons. (Before the advent of clocks the "hour" was probably of a more flexible nature.)

Where previously time's qualities had been measured by cycles, seasons, the waxing and waning of the moon, the canonical periods of prayer, and the chanting of the offices of the day, now time was quantified and reduced to numbers. But numbers can be easily arranged on a line—which mathematicians refer to as "the number line." So it was quite natural that, in place of cycles within cycles, time should also be strung out on a line and counted off in so many hours and minutes. Now instead of time cycling and returning it would stretch out indefinitely from past to future.

Time in other cultures was the god Chronos, the rotations of the gods of the Mayan calendar, the old man with his scythe, the figure calling us toward the grave. Now time was number, and number was time.

This new sense of time, based on the mechanical clock, became the standard against which other aspects of life could be measured. Events took place "as regular as clockwork." Even human beings could be clocklike. The philosopher Immanuel Kant took his daily walk with such regularity that his neighbors set their clocks by him. By the start of the nineteenth century particularly accurate clocks were called "regulators," a name that had previously been applied to certain judges and commissions. The rule of clockwork had become a metaphor for law and the good order of society. Within a clockwork universe there could be no surprises and no ambiguities, only a series of certainties strung out along the line of time.

This metaphor of the clock also applied to the heavens, as in the phrase "Newtonian clockwork." Isaac Newton had demonstrated that all motion, from the fall of an apple to the orbit of the moon around the earth, could be explained on the basis of three simple laws. With their aid it is possible to predict eclipses of the sun and moon for centuries to come. Because of this regularity, the solar system was compared to a clock, a mechanism that is stable, predictable, and understandable and which holds no irregularities or surprises.

The philosopher Wilhelm Leibniz satirized Newton's God as someone who wound up his watch at the moment of Creation and then allowed the universe to tick away by itself. Yet Newton's vision was magnificent. By stripping away the qualities of things, their taste, feel, and color, it became possible to arrive at an essence of movement—the mathematical principles of structure and transformation that underlie the material world.

Just as in the previous chapter we saw how Renaissance painters discovered the trick of linear perspective by which they could express the space and depth of the world, so too Newton produced a faithful representation of the movements of the universe in terms of number. The French mathematician Pierre Simon de Laplace claimed that if he had stood beside God at the moment of Creation he could have used Newton's laws to predict the entire future of the universe.

Laplace's fantasy exposes another aspect to this metaphor of

Newtonian clockwork. Laplace imagined himself standing beside God. This meant that he was no longer a part of the universe. Instead of being a participator within a living cosmos, he stood outside and observed its inner working in a dispassionate manner. This is also an image of Newtonian science itself. While it was possible to describe the motions of the heavens using mathematics, this "universe" turned out to be less a home in which to live than an object standing before us to be understood, described, predicted, and controlled. The values and qualities, the tastes and smells of the universe become less important, or essentially irrelevant, when compared to its mathematical description in terms of mass, position, and speed.

Newtonian clockwork also had its applications here on earth. As the moon orbits around the earth it pulls on the oceans and so produces the alternation of high and low tides. Such events, the time and height of tides, are entirely predictable, except for the minor perturbations caused by irregularities in coastline, the meeting of tidal streams in estuaries, and so on. But knowing the exact time and height of a tide is important; it even proved to be a key element in the plot of John Buchan's famous spy novel *The Thirty-nine Steps.*[1]

Newtonian clockwork appeared, at first sight, to be a perfect mechanism. There was however, a tiny grain of sand hidden deep within its wheels and cogs. When it comes to the moon's motion around the earth, or the earth's orbit around the sun, Newton's laws can be solved exactly and the appropriate numbers calculated to any accuracy desired. But what about the small additional pull of the moon on the earth as the earth orbits the sun? And what is the precise effect

[1]As a boy I remember seeing a tide predictor at the Bidston Hill Observatory near Liverpool. This great machine, the mechanical forerunner of a computer, occupied an entire room that was carefully controlled for temperature and humidity. While the moon is the driving force of the tides, other small perturbing effects, such as the shape of an estuary line or the meeting of opposing currents of water, can alter the exact height of a tide. These factors were represented in the tide predictor by a series of cogs and wheels. Through their rotations it was possible to compute tides for days and months to come. Again, certainty and predictability had become associated with regularity, clockwork, and the ability to strip away inessentials in order to describe apparently complex phenomena in terms of simple mechanical models.

of the gravitational pull of the asteroid belt on the orbit of Jupiter? These tiny effects are analogous to the perturbations that coastline irregularities have on tide predictions. Scientists call this astronomical problem the three-body problem. It asks: How do three or more bodies move under their mutual attractions of gravity?

While the two-body problem can be solved exactly, there is no simple solution to the three-body problem. No single equation can be written down directly and used to calculate numerical answers to any degree of accuracy. This does not mean that Newton's laws are incorrect or approximate. Rather, the corresponding mathematical equations present insurmountable difficulties that make it impossible for a general solution to be written down in a direct way. In the case of the simpler two-body problem, it is just a matter of inserting the numerical values for the position, speed, and mass of the earth and sun into the relevant equation, and the answer pops out. But when the mutual pulls of earth, sun, and moon act together on each other this simple approach no longer works.

Astronomers found a way around this problem using an approach called "perturbation theory." In perturbation theory you begin with the reasonable assumption that the moon's effect on the earth's orbit around the sun is very small. Start with the simple two-body problem, the earth's orbit around the sun (neglecting the moon), and then apply a small correction (called the "perturbation") to take into account the much smaller pull of the moon. To this first correction apply another, even smaller, correction. And then a third correction, and so on ad infinitum. In practice scientists don't need to add too many of these corrections because, after the first, the size of successive corrections becomes so tiny as to make no practical difference to the value of earth's orbit of the sun.

It is a case of "wheels within wheels."[2] Using perturbation theory astronomers made tiny corrections to the orbits of the planets to account for the gravitational pulls of smaller third and fourth bodies.

---

[2]The complex machinery of cogs I had seen at Bidston Observatory was the mechanical analogy of perturbation theory, in which the motion of ever smaller cogs adds in tiny corrections to the tide predictions.

The results satisfied astronomers but left mathematicians feeling uneasy. Astronomers were adding together a number of tiny corrections—admittedly each one was much smaller than the other. It is not unreasonable to assume that a few very small things add up to another small thing. But what about adding up an infinite number of very small things? How do we know that these won't sum up to something large?

Mathematicians love to play with patterns of numbers and devise ways for summing up infinite series of ever-smaller numbers. Take, for example, the series known as $1/n^2$.[3] The first member of the series is $(\frac{1}{2})^2$, that is, $\frac{1}{4}$. So imagine our unperturbed answer is 1.0000. Adding this first member of the series, which we can also think of as a "correction" to 1.0000, gives us $1 + \frac{1}{4}$, or 1.25. The next member of the series of "correction," is smaller $(\frac{1}{3})^2$, that is $\frac{1}{9}$ or 0.1111. This is now added to the first correction. The new "corrected" answer is 1.3611. The third correction is $(\frac{1}{4})^2$, or $\frac{1}{16}$, which equals 0.0625 and brings the answer to 1.4236. Additional corrections are even smaller $\frac{1}{25}$, $\frac{1}{36}$, and $\frac{1}{49}$, but there are an infinite number of them.

In this case mathematicians know the precise answer when all these terms are summed. Starting with the number 1 and adding an infinite number of corrections we arrive at the answer 1.6449. The initial answer of 1.0000 has been somewhat perturbed, but even with an infinite number of corrections it remains finite.

There are many such series where an infinite number of corrections add up to a finite answer. But what about the series: $1 + \frac{1}{2} + \frac{1}{3} + \frac{1}{4} + \frac{1}{5} + \frac{1}{6}$ and so on? Again each correction becomes smaller and smaller. However, in this case mathematicians know that when an infinite number of *these* corrections are added together the answer blows up in our face. It is infinite. This was what worried mathematicians when they used perturbation theory to solve the three-body problem. How did astronomers know that in every planetary case the effect of an infinite number of small corrections would always result in a finite correction to an orbit? What happens if these corrections blow up? What does this mean for the orbit of a planet or an asteroid?

---

[3]The superscript above and to the right of *n* indicates that the number *n* must be multiplied by itself; that is $n \times n$.

Toward the end of the nineteenth century Henri Poincaré attempted to lay this problem to rest. While he was still unable to solve the three-body problem specifically, he found a way to say something general about the overall shape and behavior of its solutions.

Poincaré showed that, in most cases, things turn out as everyone expected—small influences result in small effects, and the more accurate solutions are close to the simpler two-body solutions. Nevertheless, this need not always be the case. In some very exceptional instances the perturbed solutions "blow up." The addition of a large number of very small effects accumulates rapidly, and instead of planets being "regular as clockwork," for certain critical arrangements the system becomes unstable. In other words, Poincaré had discovered chaos hidden within the heart of the Newtonian universe. Newtonian clockwork was only regular under certain conditions. Outside this boundary, physics was faced with uncertainty.

How does this happen? The rotation of earth round the sun presents a relatively simple problem for Newtonian science. The sun pulls the earth toward it; likewise the earth exerts a pull on the sun. Now add in the effect of the moon. As the moon rotates around the earth it exerts a slight pull, which has the effect of slowing down and speeding up the earth's motion. It is alternately pushing the earth toward the sun and pulling it away again.

In the case of the earth–sun system the moon's effect is not very large. But for certain critical arrangements of other planets, "resonance" can take place. To understand resonance, think of a heavy man on a swing and a small child who gives the swing a nudge from time to time. (These nudges can be thought of as perturbations to the motion of the swing.) Over all, these nudges have little effect on the man's regular swinging back and forth. But suppose the child nudges the man each time he reaches the highest point of his swing. If each tiny nudge is timed exactly the nudges will begin to add up. In swing after swing the man goes higher and higher. This effect of a very small perturbation that accumulates from oscillation to oscillation is called resonance.

In the case of two planets in orbit around the sun the second may be nudging the first in such a way that these nudges resonate exactly with that planet's "year." In turn, the first planet also nudges the second. In this way tiny effects accumulate until the entire system acts wildly. Effects from one planet are feeding back into the orbit of the other.[4]

The same thing can happen with critical arrangements of the orbits of two planets around the sun. The very tiny perturbing effect of one on the other feeds back with each orbit of the planet, amplifying until the whole system becomes unstable. In this way, Poincaré pointed out that within one of the most basic of all certainties—that the sun will rise each morning—was hidden the potentiality for instability, surprise, uncertainty, and even chaos.

Poincaré published his result in 1900. It was the same year as Planck's hypothesis about the quantum nature of energy. Five years later Einstein's special theory of relativity appeared and then the flurry of contributions from Bohr, Sommerfeld, Heisenberg, Schrödinger, Pauli, Fermi, and Dirac that established modern quantum theory. No wonder Poincaré's remarkable result was marginalized and did not remain within the center of the scientific limelight. Physicists and mathematicians were also discouraged by the difficulties they would have to face if progress were to be made beyond Poincaré's initial result. After all, many scientists prefer to work on problems that will yield results for publication, since publication often leads to promotion!

It was not until halfway through the twentieth century that breakthroughs occurred that gave birth to the present science of chaos theory. Three Russian mathematicians, A. K. Kolmogorov, Vladimir Arnold , and J. Moser, came up with general ways to picture the sort of problems on which Poincaré had been working. Another advance was the development of computers that could solve highly complicated equations numerically and display the solutions on a screen. Today scientists and mathematicians can visualize complicated systems and "see" what the solutions look like.

---

[4]Feedback also occurs in a public address system. When the amplifier is turned up too high or the microphone is too close to a speaker, a tiny noise in the hall is picked up by the microphone, amplified, and emitted out of the loudspeakers. In turn, the louder sound from the speaker is picked up by the microphone, amplified again, and sent out into the hall. As an initially quiet sound circulates it *feeds back* from microphone to speaker until it grows to an ear-splitting screech.

Kolmogorov, Arnold, and Moser (KAM) confirmed that chaos and stability could exist within the solar system. In most situations the orbits of the planets remain stable for tens of millions of years, but for certain critical arrangements of orbits, the tiny pull of one planet on another planet accumulates and "feeds back" to the first planet. This may be the explanation for the gaps in Saturn's rings. Calculations show that if a rock were placed in one of these gaps its orbit would become so erratic that it would fly off into outer space or collide with material in the other rings. A similar explanation may also account for the absence of a planet lying between Mars and Jupiter. Matter that attempted to coalesce in this region may then have been subject to erratic forces. As a result, instead of forming a planet, it gave rise to the asteroid belt, a collection of rocks and mini planets.

KAM's approach, along with high-speed computers, applied not only to the solar system but also to a host of other situations, including weather, water waves, the stock market, fluctuation in the size of insect populations, the spreading of cracks and faults in metals, traffic patterns, brain activity, heart beats, prison riots, the mutual interaction of certain chemicals, and turbulence in a pan of heated water. Today mathematicians, engineers, physicists, chemists, scientists, biologists, environmentalists, economists, sociologists, and even psychotherapists use ideas from chaos theory and work on ever more complex systems to the point where they find themselves joined by artists, designers, animators, filmmakers, composers, and computer hackers.

A variety of names is associated with this new science: nonlinear systems theory, catastrophe theory, chaos theory, complexity theory, self-organizing systems, open systems, general systems theory, fractals, strange attractors, far-from-equilibrium systems, autopoiesis, and so on. In popular accounts they all tend to be lumped together under the general rubric of chaos theory. As with quantum theory, chaos theory places strict limits on certainty. It indicates that we must always be willing to accept some degree of "missing information."

But what exactly is chaos? Take something as simple as a pan of water on the stove. Water at the bottom of the pan begins to heat and, being warmer, it is less dense and therefore tends to rise. Water at the top of the pan, at room temperature, is heavier and begins to fall. Warm rising water therefore fights for space against cooler descending water. Inevitably the result is chaos—a complex series of competing flows within the pan to the point where it seems impossible to predict how the water will behave from region to region. Similar forms of turbulence occur in a host of different systems: when winds encounter city skyscrapers, as speedboats rush across lakes, or when commuters entering a subway station must fight an exiting crowd.

Machines that vibrate out of control, static produced in electronic devices, rivers in flood, atmospheric storms, fluctuations in the stock market, and fibrillation of the heart are all examples of systems that appear unpredictable and out of control, systems in which what happens from moment to moment appears to be a matter of pure chance rather than of scientific law.

Until KAM and high-speed computers came along such chaotic systems were regarded as too messy to be within the province of science. Theoretical physicists and engineers preferred not to think about them. If you push a steam engine or an automobile too fast it begins to shudder and quake to the point where it may self-destruct. Such behavior is to be avoided rather than made the subject of research. And if you adjust the settings on an amplifier and the room is filled with static then clearly you have a badly designed amplifier.

Today all such systems are open for study using the approach known as chaos theory. And if scientists have given up hope of ever fully describing a chaotic system, at least they have come a long way toward understanding them.

Chaotic Systems

Scientists no longer throw up their hands in horror at chaotic systems for they know that such systems conceal many interesting secrets. Chaos itself is one form of a wide range of behavior that extends from simple regular order to systems of incredible complexity. And just as a smoothly operating machine can become chaotic when pushed too hard (chaos out of order), it also turns out that chaotic systems can give birth to regular, ordered behavior (order out of chaos). Chaos theory explains the ways in which natural and social systems organize themselves into stable entities that have the ability to resist small disturbances and perturbations. It also shows that when you push such a system too far it becomes balanced on a metaphoric knife-edge. Step back and it remains stable; give it the slightest nudge and it will move into a radically new form of behavior such as chaos.

All these systems exhibit what is called nonlinear behavior. Nonlinear systems behave in rich and varied ways. In a linear system a tiny push produces a small effect, so that cause and effect are always proportional to each other. If one plotted on a graph the cause against the effect, the result would be a straight line. In nonlinear systems, however, a small push may produce a small effect, a slightly larger push produces a proportionately larger effect, but increase that push by a hair's breadth and suddenly the system does something radically different. Put gentle pressure on the accelerator pedal of your car and the speed increases. The greater the pressure, the faster the car goes. This is linear behavior. But when the accelerator pedal is pressed to its limit, the passing gear kicks in and the car jumps forward in a nonlinear way. In the case of three astronomical bodies all those tiny pushes and pulls on the orbits can feed back into each other and, when resonance occurs, accumulate into a much larger overall effect.

Over a limited and fixed range of behavior, external influences can have a predictable effect on a nonlinear system. But when the system reaches a critical point, a knife-edge called a "bifurcation point," it will jump in one of several different directions, often in an unpredictable way. Put a ball bearing at the bottom of a bowl and a small push will send it a little way up the side until it falls back again. But balance it on the lip of the bowl and a single breath of wind will cause it to fall back into the bowl or alternatively fall onto the floor and roll away into the corner of the room.

A system at a bifurcation point, when pushed slightly, may begin to oscillate. Or the system may flutter around for a time and then revert to its normal, stable behavior. Or, alternatively it may move into chaos. Knowing a system within one range of circumstances may offer no clue as to how it will react in others. Nonlinear systems always hold surprises.

What is of particular relevance to the argument of this book is that such systems are also discovered in human organizations, the stock market, traffic patterns, spread of diseases, fluctuations in population size, and so on. In all these cases, and many more, a tension exists between what can be known and determined for sure, and what lies beyond our predictive capacity.

Chaotic Populations

Planetary chaos was introduced through the metaphor of grit in the Newtonian clock. There are other examples where regular, cyclical behavior conceals the seeds of chaos. Take as an example of regular behavior a reedy lake containing trout and pike. If the pike are too rapacious they will consume their source of food and start to die out. But there are always a few trout hiding in the reeds and, freed from the threat of so many predator pike, their numbers increase. Soon the lake is well stocked with trout and the few remaining pike discover they can have a field day. Pretty soon the pike population increases and many of the trout get eaten. Now the cycle begins again. Hungry pike discover that their prey cannot be easily found and so they begin to die of starvation. Year after year, and generation after generation, the number of trout and the number of pike oscillate up and down in a stable and predictable way.

Cyclical oscillations of predator and prey look very much like the ticking of a clock. But in this case the origin of the pendulum swing is not mechanical; rather the results of one cycle feed back into the next in a repetitive way. Mathematicians call this form of repetition "iteration." Iteration means that the output of one calculation, or of one cycle, is the input for the next. Some iterations lead to stable situations, such as the population of pike and trout, while others produce chaos.

To see how chaos can emerge out of regular population cycles, let us change the example slightly. Instead of pike and trout we'll take rabbits. Release a pair of rabbits on a virgin continent and they will breed until they have spread over the entire land. But suppose these rabbits arrive on a small desert island. At first they breed and multiply, but soon they are eating the vegetation as fast as it can grow. Like the pike that eat too many trout, the rabbits begin to die out.

There are two competing factors at work on the population, one causing it to expand by breeding and the other causing it to die off because of limited space in which to live and limited food to eat. Like the example of the pike and trout, population size is determined by an iterative situation because young rabbits of one season become the breeding pairs of the next. It turns out that the mathematical equation that models this behavior is quite simple and, provided you put in a value for the birth rate, the population can be predicted for years and years to come.

To explore this example even further we must now forget about real rabbits and deal with hypothetical computer rabbits whose birthrate can be adjusted as we choose. Real rabbits don't work in this way, unless they are given hormones, but the example itself applies to a host of other real-life situations, from the spread of rumors and the distribution of genes in a population to certain chemical reactions and insect damage to crops.

With a low birthrate the initial pair of "rabbits" breeds and the population increases until it reaches a stable level that remains static from generation to generation. This population is exactly in balance with the resources of the island. It is the stable sustainable population for that particular environment.

With a higher birthrate the population increases more rapidly, temporarily overpopulating, then falling again, and after a time rising again. The result is a stable oscillation in population size, a predictable sequence of fat and lean years that exactly mirrors the behavior of pike and trout in a lake.

But suppose the birthrate is higher still. The result of a mathematical analysis shows that within the first oscillation of fat and lean years can be found a second oscillation, a subcycle, a cycle within a cycle. It now takes four turns of the cycle to come back to the starting point.

Increasing the birthrate even further means that new oscillations are added. Now it is a case of wheels within wheels within wheels, or oscillations within oscillations within oscillations. The situation becomes increasingly complex; population scientists would have to gather data from many years to work out the complex pattern of oscillations and so be able to predict the population for the following years.

While the cycles within cycles are complex, the underlying mathematical equation is quite simple, and with the help of a computer, scientists can watch the way cycles increase in complexity each time the birthrate is increased very slightly. It would be natural to assume that the end result would be an infinite number of cycles, a vast machine of incredible proportions containing a limitless number of cogs within cogs. But, infinitely complex as this may be, this is still a regular form of behavior, since if one were willing to wait for an infinitely long period of time the same behavior would cycle round again.

This, in fact, is not what happens. The system reaches a critical point at which the very slightest increase in birthrate no longer generates an additional cycle but, rather, chaotic behavior. The population now jumps at random from moment to moment. No amount of data collection can be used to predict the population at the next instant. It appears to be entirely without order. It is truly random and chaotic.

With the aid of this example we encounter one of the paradoxes that lies within the heart of chaos theory: What does it mean to say that something is random or that it has no order? Toss a quarter in the air and you can't predict if it will come down heads or tails. Throw a ball into a roulette wheel and you don't know if it will end on black or red. The result is random. Knowing the results of a long sequence of coin tosses is no help in predicting the next result. If someone has tossed six heads in a row the chance of the next throw coming down heads remains exactly 50:50. The sequence of heads and tails is random. But this does not mean that the process by which the coin lands head or tails is itself without any order. Each time you flick a coin you use a slightly different amount of force and so the coin spins in the air for a slightly different amount of time. During this same period it is buffeted by chance air currents and when it lands on a table it bounces and spins before settling down heads or tails. In coin tossing the coin is subject to a large number of perturbations and disturbances that are beyond our control. Moreover these contingencies are so complex as to be beyond any normal sort of calculation. Nevertheless, at every instant of the coin's flight everything is completely deterministic.

The same thing happens in a pinball machine or a roulette wheel. In both cases the ball is buffeted in a complex series of collisions. Rather than there being no order there is an order so complex as to be beyond prediction.

Likewise, while chaos theory deals in regions of randomness and chance, its equations are entirely deterministic. Plug in the relevant numbers and out comes the answer. In principle at least, dealing with a chaotic system is no different from predicting the fall of an apple or sending a rocket to the moon. In each case deterministic laws govern the system. This is where the chance of chaos differs from the chance that is inherent in quantum theory. When a rock rolls down a mountainside it is knocked here and there by the contours of the slope. The end result, where it finally lands, is random. Yet each individual bump is totally deterministic and obeys Newton's laws of motion. Its extreme complexity arises out of the huge number of external perturbations acting on the rock. Chaos and chance don't mean the absence of law and order, but rather the presence of order so complex that it lies beyond our abilities to grasp and describe it.

Chance isn't always caused by external perturbations. In a fast-flowing river an individual, tiny region of water is being pushed this way and that by the river itself, like a rock rolling down a mountainside. The entire system acts as a complex series of internal perturbations pushing and pulling on each aspect of itself. Feedback and iteration act within the system to create ever greater complexity. The result is a turbulent river.

Now let us take another glance into the mirror of chaos and discover that what we take to be a random system without any visible order can also be seen as an order of infinite richness and complexity. Each aspect of a chaotic system is deterministic and governed by internal feedback, constant iteration, or complex external perturbations. Similarly a computer calculation must simulate these physical effects. Working with a turbulent river or a population with a varying birthrate means that the result of one stage of the calculation (its output) becomes the input for the next round of calculations. Like the system itself the computer calculation iterates itself again and again, with each output being the input of the next cycle.

This brings us to yet another paradox of chaos, for, although the equations of a system are totally deterministic, the final results can never be calculated exactly. Even the fastest and largest computers are finite. They may carry out a calculation to ten decimal places, which is good enough for most purposes. But this means that there is always uncertainty in the final decimal place—one part in ten billion. This seems unimportant until one realizes that this tiny uncertainty is being iterated around and around in the calculation. Under critical conditions the cycling of even an almost vanishingly small uncertainty begins to grow until it can dominate the entire result.

The meteorologist Edward Lorenz discovered this in 1960 when he was iterating some simple equations used in weather prediction. To speed up the calculation he dropped some decimal places and, when the calculation was finished, to his great surprise he discovered that the resulting weather prediction was vastly different from his initial, more accurate calculation. A small uncertainty in the initial data of the weather system had swamped the final calculation.

By analogy to the computer calculation, when a real weather system is in an unstable condition a small perturbation can produce a radically different change of weather. With a system balanced on a knife-edge or at a bifurcation point, even the flapping of a butterfly's wing can send it in a totally different direction. The ancient Chinese drew attention to the interconnectedness of all things by saying that the flapping of a butterfly's wings can change events on the other side of the world. In chaos theory this "butterfly effect" highlights the extreme sensitivity of nonlinear systems at their bifurcation points. There the slightest perturbation can push them into chaos, or into some quite different form of ordered behavior. Because we can never have total information or work to an infinite number of decimal places, there will always be a tiny level of uncertainty that can magnify to the point where it begins to dominate the system. It is for this reason that chaos theory reminds us that uncertainty can always subvert our attempts to encompass the cosmos with our schemes and mathematical reasoning.

There is yet another reason why a system, deterministic in principle, can be unpredictable in practice. Leaving the limitations of computers aside, it is impossible to collect all the data needed to characterize a system exhaustively; that is, without any degree of error or uncertainty creeping in. In turn, that uncertainty rapidly blows up when systems iterate within themselves. Take the world's weather again. A mathematical argument (based on the properties of fractals) demonstrates that there can never be a sufficient number of weather stations to collect all the information needed to describe the fine details of the weather at any one time. (The fractal dimension of weather is larger than the fractal dimension of any network of weather stations.) While it is possible to predict weather trends for days in advance we can predict exactly what the weather will be at any precise instant. It may look like heavy rain tomorrow but we cannot know just how many millimeters of rain will fall at a particular spot, or the exact time that rain will begin.

In addition, just as in quantum theory, the very act of observation of a system disturbs the properties of the system: the effect of introducing a probe or making a measurement when a system is at a bifurcation point or in a chaotic state can also cause that system to respond in an unpredictable way. Although it is always possible to adjust and fine-tune a linear system, things are entirely different when it comes to nonlinearity. In certain regions of behavior the system may respond to a corrective manipulation; in other regions a small correction may push the system in an unexpected direction.

Chaos, Chaos Everywhere

The previous chapter argued that the way we represent the world has a deep influence on what we see. Chaos theory provides an excellent contemporary example of this phenomenon. Today we tend to "see" the world, ourselves, and our organizations in terms of attractors, chaos, self-organization, and the butterfly effect. Economists and financial analysts look for patterns of self-similarity within the daily and hourly fluctuations of the stock market. Therapists speak of strange attractors governing the repetitive behavior of their clients. Community leaders and business consultants are concerned with the dynamics of self-organizing systems. Moviemakers create planetary geographies using fractal generators. Suddenly chaos, complexity, and self-organization surround us to the point where the general public is using terms more generally associated with mathematicians and theoretical physicists, whereas just half a century ago, no one had ever heard of such terms.

Only a few decades ago the fluctuations of the stock market were seen as purely random. Organizations and business were studied in terms of rules and hierarchies and good and bad managers. And "chaos" itself? It was simply a word used to mean a pattern without any order, an aberration, something not worth studying or taking seriously. Chaos was the garbage can into which everything was thrown that could not be represented by means of simple rules and behaviors. And what we now know as fractal orders were once called by mathematicians "a gallery of monsters."

How did such a striking change in attitude come about? Why did people begin to take an interest in chaos and notice strange, new, complex patterns of order in what they had previously taken as random events? Again, the short answer is that we mainly see what we already know. Or to put it another way, we could only begin to "see" the inner world of chaos once we had discovered ways of representing it. Once we are given a mental map of the world of chaos we can begin to discern its features.

The development of high-speed computers and new mathematical approaches made it possible to describe the general nature of chaotic systems, apparently random fluctuations, and highly complex patterns. These features of nature had always been present, but until the means to represent them had been discovered they were essentially invisible to us. These very important aspects of the world had been ignored because we had no real way of looking at them. In 1900 we saw a world of law, order, and certainty in which chance and randomness were unwanted exceptions. Today uncertainty and chaos are seen as essential to the hidden order of the cosmos.

Chance

For the past few hundred years Western science, and the Western mind, have been preoccupied with notions of certainty, predictive power, and the exercise of control. Other societies are willing to accept flux and uncertainty. They live in the Tao, within the flow of things, and tolerate the fact that they will never know all there is to know about the universe. By contrast, the Western mind has been seeking a story with a definite ending. Science wants theories that are finite and rounded off. A good theory should not leave gaps, areas of ambiguity, or uncertainty. Moreover, as in some Freudian death wish, physics seeks to bring about its own end. It desires the ultimate answer, the "theory of everything" that will bring closure to its activities. With the ultimate equation, theory will be finished, all questions will be answered. We will know once and for all the story of the universe. In fact that term, "the universe story," has been used by Thomas Berry and Brian Swimm as the title of a book and a project to provide a contemporary scientific account of the universe of similar mythic proportions to that of Dante in the Middle Ages.

Most societies have their origin stories, ways of linking their present world and society to the creation figures of the past. Some stories concern the creation of the world. But often the world is already present as given and the stories are about the naming of things, the origin of medicine, language, cooking, and writing. Berry and Swimm intend something similar with their *Universe Story.* Yet traditional origin stories have an open quality to them or involve the role of clowns and tricksters such as Coyote, Raven, or Brer Rabbit who turn things upside down and subvert the order of the world.

Until the twentieth century forced us to face the basic uncertainty of the universe, we asked science to present us with comfortable bedtime stories, ones in which "everything comes out all right at the end." Science believed in the parsimony of the universe and applied Occam's razor.[5] There could only be one right theory and every choice should be judged as being good or bad. Now chaos theory is telling us that if we desire total certainty, if we want to hold the universe in the palm of our hands, we have to leave the human race behind and become godlike beings who can observe and measure a system without in any way

---

[5]Occam's razor states that in the face of alternative explanations one should accept the explanation that is simplest and most direct.

disturbing it. As in Laplace's fantasy of being present at the creation of the universe, such beings are omniscient to the point where they can gather complete and total information about a system. They possess computers of infinite power, computers the size of the universe itself, that enable them to understand the inner workings of that same universe.

But we are finite creatures. Total knowledge and predictive power will always be beyond us. We have to accept that we can never know the universe fully and totally. We must learn to live with a measure of uncertainty, paradox, and ambiguity. We must acknowledge that vital pieces of information may always be missing. That is the price we pay for entering fully into the life of the cosmos, for becoming participators in nature instead of mere observers. Living in the universe gives us obligations and responsibilities. Each of our acts of observation will in some way disturb the universe and we must accept full responsibility for the consequences of these actions.

Feeling Out Trends

This does not mean that we must wash our hands of chaotic systems. While their fine details remain forever beyond us we may still be able to detect patterns within their behavior that are not totally random. Stable systems, such as predator and prey (see the earlier example of pike and trout), are in the grip of what scientists call an attractor. Just as a magnet attracts iron filings into a fixed pattern, so the attractor of a complex system pulls its dynamics, or behavior, into characteristic repetitive directions. Perturb the system and its attractor pulls it back on track. Attractors are a little like Jungian archetypes, always acting in the background. If a person is in the grip of a particular archetype—the hero, the *puer aeternis* (eternal golden youth), the devouring mother—this will influence the pattern of behavior within relationships, work, and so on. Likewise knowing the shape of an underlying attractor helps us to predict what a system's behavior will be.

Just as an attractor governs a stable system, so a chaotic system is governed by what is called a strange attractor. This means that, although behavior may on the surface appear totally chaotic and infinitely complex, it nevertheless originates from an underlying pattern, for the strange attractor itself has an underlying fractal structure. Fractals are complex patterns in which a particular element of the pattern is repeated at ever decreasing scales ad infinitum. Likewise, while the behavior of a system in the grip of a strange attractor is chaotic, varying unpredictably from moment to moment, these jumps in behavior mirror each other at ever decreasing scale and take place within a certain zone, or range, of possibilities.

Economists have compared the behavior of the stock market to a system in the grip of a strange attractor. While there are overall trends that indicate which stocks are going to rise over the next weeks and which will fall, within these trends can be found fluctuations that, at first glance, appear random. Yet the "random" fluctuations that occur over say, one hour, mimic similar random fluctuations over a day, and over a week. Mathematicians call this self-similar behavior. A fractal displays similar patterns at ever decreasing scales, likewise small fluctuations within the stock market have a fractal structure, and while remaining unpredictable in their fine details, the overall patterns are imitated at smaller and smaller time intervals.

Although the detailed moment-to-moment behavior of a chaotic system cannot be predicted, the overall pattern of its "random" fluctuations may be similar from scale to scale. Likewise, while the fine details of a chaotic system cannot be predicted one can know a little bit about the range of its "random" fluctuation.

Intermittency

Up to now we have looked at systems in which simple order breaks down, or disappears, into that highly complex swirl of behavior called chaos. Yet the theory of nonlinear systems presents us with a paradox, for behind the door marked "chaos" lies a world of order, and behind that door marked "order" can be discovered chaos.

Let us return to the sudden burst of noise from an electronic apparatus—an amplifier connected to a loudspeaker perhaps. Electronic engineers know of a problem called intermittency. This occurs when the regular, ordered output of an amplifier is suddenly swamped by random "noise." These periods of random noise can also cease suddenly and give way to periods of regular behavior. When intermittency is occurring we have the alternation of randomness with simple order.

It would be easy to say that a defect in the design of the amplifier (in fact a nonlinear amplifier) results in the occasional breakdown of regular behavior to produce chaos. On the other hand, it is equally true to say that periods of chaos (highly complex behavior) break down to leave regular behavior. In one case chaos emerges out of simple order, in the other order emerges out of chaos.

Human societies have their periods of chaos—Carnival, Mardi Gras, Oktoberfest—in which normal social rules are abandoned. Men dress in women's clothing, married people indulge in sexual license, there are orgies of eating and drinking, night is turned into day, authority is mocked, and the Fool rules the day. This can be seen as a temporary breakdown of the stable order of society and the lapse of rule. On the other hand it could be that within the apparent chaos of the carnival can be found the source of a society's order over the rest of the year.

Self-Organization

In some cases chaos rules when order is relaxed, in others order has its seeds in the realm of chaos. Go back to that example of a heated pan of water. Competition between hot water rising from the bottom and cooler water descending from the surface produces haphazard, chaotic behavior. But with the right degree of heating these apparently random flows and counterflows suddenly settle down and organize themselves into a regular pattern of hexagonal cells of rising and falling water. This pattern remains stable, provided that there is a constant flow of energy, as heat, through the system. Similar patterns are found in deserts, where the competing flow of hot air rising from the sand meets cooler air falling from above. The result is that regular patterns of rising and falling air move grains of sand until hexagonal patterns form on the desert floor, just like the cells in a bee hive.

The cells in a heated pan of water, or the movement of sand in a desert, are examples of order arising out of chaos. They all occur in what scientists call *open systems.* When energy flows through a system, such as heat in a pan of water, the system can order itself into a stable structure.

A river provides another example of what is termed self-organization. During the summer it flows slowly with hardly a ripple to disturb its surface. Where there is a rock in the river the water divides and flows gently past the disturbance. But once the spring rains arrive, the river flows faster. In many ways the movement of particular regions of water appears chaotic and turbulent, but notice what happens as fast-flowing water encounters a rock. Now a vortex appears downstream from the rock. It is a stable form that has emerged out of the chaotic order. These vortices are remarkably stable. Throw in a stone and the vortex may be disturbed for only a moment, but then continues as before.

A vortex is an example of the way an open system organizes itself to produce a stable structure. Unlike the pan of water, in which an energy flow produced stable patterns, this time it is matter—water—that is flowing through the vortex. As long as the river is in spate, this structure is remarkably stable. As soon as the flow subsides, the vortex disappears.

Natural and social open systems exhibit many examples of self-organization, systems in which regular behavior and stable structures emerge out of chaos. These are found in everything from traffic flows, economic systems, the movements of goods and services, to certain types of waves in canals and rivers and even Jupiter's giant Red Spot. Some, like the vortex, are open to a flow of matter, others to a flow of energy, or even information.

A city can be thought of as a self-organized system that has structured itself over a historical period. It maintains its form by virtue of a complex network of flows—money, food, energy, people, and information. Provided that these flows are maintained at a certain level, the city will sustain itself, garbage will be moved, people will have enough to eat, taxes will be paid, and social services will function. But if any one of these flows should be interrupted for a long enough period, the city would collapse and chaos would reign.

Again one of the powerful lessons of this book is being repeated for us. That is, our acceptance of a degree of uncertainty is the very essence of being alive in the universe. Many systems in nature and human society have evolved through processes of self-organization. They were not put together in a mechanical way, by bringing various parts together and arranging them according to some hierarchical scheme and overarching law. Rather they emerged through the interlocking of feedback loops and out of flows to and from the external environment. In this sense, the stabilities of our lives, of our organizations and our social structures, do not arise out of fundamental certainties but from out of the womb of chance, chaos, and openness. Patterns in a pan of heated water and the vortex in a river are particularly simple examples of order emerging out of chaos. Likewise human society itself, with its cities, international governments, and global economics can only exist through this dynamical dance between chaos and order.

The open systems that fall under the umbrella of chaos theory have a large number of components that interact together and engage in mutual feedback. Traditionally, physicists preferred to study isolated systems where all conditions could be carefully controlled. Such systems behave in regular ways and contain no surprises, so that carefully controlled experiments always match the predictions of theory. But today we realize that nature's open systems are far richer and more interesting. Their behavior is a product of their ability to organize themselves and respond in varied ways to changing environments. It is only relatively recently, because of the long-standing theoretical difficulties involved, that such systems have begun to be studied in a systematic way.

This contrast between the versatility and flexibility of self-organization and the behavior of mechanical systems can be illustrated by comparing life in a village to that within a traditional army. To function effectively during war, an army must have a predetermined and well-understood hierarchy of soldiers, noncommissioned officers, officers, and so on up to the general staff of generals and field marshals. Recruits are put through a rigorous training and drill that teaches them to obey orders without question and to carry out tasks in a repetitive way. As soldiers they can be slotted in, like cogs in a wheel, so that, as with any smooth-running machine, they function efficiently as replaceable units. Officers are trained both to obey and to give orders and, in certain situations, to show initiative.

Each person entering the army fits neatly into a particular slot, so that during battles and campaigns the army machine continues to function despite a turnover in personnel. This also means that, with the exception of the highest ranks or individual acts of heroism, the skills and personalities of any particular individual have little significance. Soldiers fit into the army, rather than the army accommodating them.

Compare this with the village of Pari, Italy, where I now live. Here, while everyone has skills in common, idiosyncrasies of personality are important. Although there is a village association, often plans and decisions are made in the evening as people sit together chatting in the square, or as they stop and gossip while walking around the village. Sometimes a village meeting is called and resolutions are voted on. In other situations people simply turn up to help when assistance is needed. Over hundreds of years, and out of necessity, the village has learned to organize itself in a way that maintains its traditions and respects people for the particular skills they bring.

Whereas, in the army, soldiers are forced to sacrifice a measure of their personal freedom so as to fit in and obey, within the village a wide range of behavior, even verging on the eccentric, can be tolerated. The former type of structure is a metaphor for mechanical, hierarchical organization; the latter stands for the self-organization seen in many natural and social systems.

It is even possible to see such behavior in physics, as in the case of plasma vibrations in a metal. Back in the 1940s the physicist David Bohm was working on the theories of the plasmas, that curious "fourth state of matter" as distinct from a solid, liquid, or gas. Plasmas occur in the upper atmosphere and the corona of the sun as well as within metals. They are composed of electrically charged particles—positively charged nuclei and negatively charged electrons—and their mutual attraction and repulsion give the plasma its special properties.

When he was working on plasmas, Bohm was struck by the way they formed an electrical shield almost as if to protect themselves when an electrical probe was introduced. It was as if they were living organisms, he thought. At the same time that he was puzzling over their behavior, he was thinking about the future of American society. He knew that America was founded on a strong sense of individualism and personal freedom, but he was also concerned about how the good of the collective could be maintained. Did people have to sacrifice their individual freedom for the good of the whole? How was it possible to have free individuals and at the same time put the good of society first?

David Bohm realized that the two systems—the plasma and human society—illuminated each other. In physics he could treat the plasma in two mathematical ways. In one, he dealt with an undisciplined crowd of individual electrons. In the other he treated the plasma as a single entity, a sort of vibrating cloud. As Bohm studied the problem he discovered that, mathematically speaking, each of these descriptions is enfolded within the other. The collective behavior of the vibrating cloud unfolds out of the individual motion of the free electrons. Likewise, individual motion unfolds out of collective vibrations. But this mutual unfolding introduces a subtly different slant on the nature of individuality. The electrical charges on electrons cause them to affect each other at long distances. But the collective aspect—the vibrating plasma cloud—modifies or shields out the long-range electrical forces that operate between individual electrons. The result is that, within the ambience of the plasma, individual electrons act as if they only experience electrical forces when other electrons are very close to them. Because each individual electron contributes to the whole plasma these individual electrons are ever freer.

Bohm concluded that hidden within the apparently chaotic motion of individual electrons could be found the collective vibrations of the whole plasma. Conversely, concealed within the vibrations of the plasma are the motions of free electrons. Likewise, within a human society each individual makes free choices that in some small way may change the course of that society. Conversely, the choices we make are influenced by the meanings we find in life, and very often these meanings are the product of the society in which we live. Thus the freedom of individual choice is enfolded within the whole of society, and the meaning of that society can be discovered within each individual.

While chaos theory is, in the last analysis, no more than a metaphor for human society, it can be a valuable metaphor. It makes us sensitive to the types of organizations we create and the way we deal with the situations that surround us.

Organizing Self-Organization

It is a major leap from the simple examples of the vortex in a river and patterns in a heated pan of water to the New York Stock Exchange and the growth of the Internet. While the latter examples do have features in common with the former they are vastly more complex in their internal structure and range of behavior. Indeed, when it comes to socially based systems we reach the limits of the more simplistic metaphors of chaos theory. Such systems involve a delicate balance of dynamical structures that involve feedback loops at many levels. Their internal complexity allows them to remain open to the contingencies of the external world while maintaining internal stability.

Take, as a starting point in increasing complexity, a single living cell. To preserve their internal chemistry, cells have evolved a semipermeable membrane. This membrane allows nutrients to enter and metabolic waste products to leave. At the level of these exchanges, the cell is open to its environment, yet at the same time the stability of its internal chemistry is also being isolated from the outside world. To survive and divide, a single cell must be sufficiently open to a two-way traffic with its environment, yet at the same time it must shield its internal structure from undesired fluctuations in that same external environment.

The human body is even more complex. Collections of cells have gathered to form organs and, in turn, organs make up the body itself. The body displays a rich hierarchical structure that is maintained through the interaction of its many feedback loops involving the blood stream, nervous system, immune system, and flows of hormones and other chemicals.

The human body must be open to its environment. It scans the horizon for food. It seeks a mate. It avoids danger and eliminates waste.

And while it is looking outward it must also preserve its internal environment. From day to night, winter to summer, the body must maintain a stable core temperature. It must monitor and control levels of sodium, potassium, oxygen, and carbon dioxide in the blood. In this way a complex web of interactions maintains the activities of the brain, circulatory systems, waste elimination, and so on. Clearly with this level of organizational complexity, the human body and its functioning are a far cry from patterns in a pan of heated water.

Through the long processes of evolution, the human body developed highly sophisticated control mechanisms to maintain a high degree of internal stability (homeostasis) within a contingent world. Shift core temperature by a few degrees and the result is coma or death by hypothermia. While the message of chaos theory is that natural and social systems can self-organize out of underlying chaos, the more sophisticated the resulting system, the more a balance must be maintained between chaos and order and the more complex (and robust) must be its internal structures and control mechanisms. At one level the body appears to function in a hierarchical fashion, with its particular functions designated to semiautonomous players such as the immune system, brain, and circulatory system. At the same time, all these players are richly interconnected through a wide variety of feedback loops.

Yet despite, or indeed because of, its stability, chaos also plays a role within the body's structures and processes. Take the human heartbeat, for example. When it is totally regular this indicates disease or even the onset of a heart attack. If it demonstrates too much chaos then it has entered a state of fibrillation, and death may ensue. Instead the healthy heart maintains small ("chaotic") fluctuations around its steady beat. Good health therefore depends on allowing a small amount of chaos into the system. Something similar applies to brain patterns. When they are totally regular and free from fluctuations this indicates that a person is in coma or under a deep anesthetic. A beating heart and a functioning brain are complex systems resulting from the cooperative behavior of many smaller subsystems. In this sense, the brain and heart are self-organized systems that, for their continued health, must combine an overall goal (a regular beating heart, for example) with a measure of individuality (fluctuations within the regular beats).

The same applies to human behavior itself. We often think of "madness" as being irrational and without any order. But generally the opposite is true. Those who suffer from severe mental illness, psychosis, and so on often have restricted and repetitive behavior. An obsessive compulsive, for example, cannot tolerate the least degree of uncertainty in the environment and so such people engage in elaborate repetitive rituals such as arranging the objects in their room and touching each one in turn. By contrast, those who are mentally healthy are capable of a wide range of responses and forms of behavior. They can adjust to changes in their environment, tolerate ambiguity and uncertainty, take intuitive leaps, and make plans even when they do not have full information as to a particular situation.

Organizing Organizations

The individual who knows many things is more likely to survive and prosper in today's rapidly changing world than the highly specialized expert who has restricted his or her knowledge to one skill alone. It is possible, for example, to design a system in a rigid way so as to protect its inner functioning. Provided the environment is stable, such systems can survive indefinitely. Evolution is full of such examples of insects, plants, and animals that have evolved to fill a particular ecological niche within an unchanging environment. For them, fluctuation, chaos, and change would present a real danger. Others, including the rat and the human being, are able to exploit change and uncertainty to their own advantage.

The same is true of human organizations. Some businesses have evolved to do one thing extremely well and to go on doing it in response to a relatively constant demand. It would make no sense to introduce sweeping changes or a new range of products in some circumstances. Yet, when the environment in which these businesses operate undergoes a major change, they will die out and be replaced by something entirely different. Dynamically changing environments, which include many of the social and economic environments of our present world, demand social organizations that are sufficiently flexible to adjust to unforeseen fluctuations, to adapt to the unknown and be willing to exploit new pathways and strategies as circumstances change.

Chaos theory cautions us that complete knowledge and control will always elude us. Nevertheless, just as the human body must retain a measure of homeostasis when all around is changing, so too a business cannot operate through total unpredictability, chance, and contingency. While it may be open to change, a business must also make a profit, or at least avoid heavy losses, even when the market is unstable. Economists need to know the effects of changes in the bank rate. Governments have to make policies for years ahead. How then can organizations function effectively while at the same time tolerating a measure of ambiguity and uncertainty in the world around them? The answer is that a measure of flexibility and what perhaps could be called biodiversity is required.

Chaos theory invites us to reflect upon the structures and organizations that surround us, from our workplace to the community in which we live, our golf club, religious organization, school, and even the national government, the United Nations, and multinational corporations. How do these organizations function? Do they appear rigid and hierarchical? Can they tolerate a degree of uncertainty? Are they able to respond to the needs of individuals? How easy is communication within and between the different levels of the organization? Are suggestions appreciated and acted on? Is the image the organization presents to the outside world different from that seen by its employees? How rich are its feedback loops? How complex is its internal structure? How flexible is it to degrees of unpredictability?

The structures of organizations are always present in both explicit and implicit ways. When a corporation occupies a high-rise, its structure is quite obvious. Directors and managers occupy the upper floors. They have their own individual offices, washrooms, and dining room. Those on the floors below work in open-plan offices and use a cafeteria. They are clearly lower on the pecking order.

Sometimes a physical building expresses the essence of an organization, the face it wishes to present to the world. In other cases it is something left over from an earlier period of the organization's history that no one has taken the trouble to change. But in all cases, physical surroundings have a subtle effect on those who work in the building. For example, what role is played by that oil painting behind the vast mahogany desk? Is it there to impress clients? or to bolster the ego of the director? Mussolini positioned his desk at the end of an extremely long room so that each person summoned by the dictator became diminished as he or she was forced to walk that long distance under the scrutiny of Il Duce's eyes. By contrast the highly respected Canadian politician Mitchell Sharp, when he was minister of External Affairs, chose to queue up and eat in the staff cafeteria along with everyone else. He not only represented a democracy but practiced the spirit of this democracy in his daily life

And how is creativity encouraged and used within an organization? How rich are the lines of communication and feedback between individuals and the various sections of the organization? What level of initiative do people have? Are the rooms and open-plan offices anonymous? or do they express the personality of each occupant? Do the employees feel that they are only carrying out the tasks that have been assigned to them? or are they contributing something essential of themselves? Are they bringing their own particular skills and life experience to the organization? Are they being respected both as persons and as skilled workers? In short, are they engaged creatively so that they feel a deep satisfaction by the end of each week's work?

I was once walking around a research organization with a scientist from the Massachusetts Institute of Technology (MIT). One of the researchers at the bench asked why his own organization had not produced scientific work of comparable status to that of the Boston research institute. The MIT scientist replied, "That's easy to understand. I drove past your labs at seven o'clock last night and all the lights were out. At MIT the lights are still blazing after ten o'clock!" One organization was offering challenges and personal engagement; the other was presenting routine.[6]

---

[6]Of course a very different interpretation is possible for the same phenomenon: that one organization was full of ambitious workaholics and the other staffed with individuals who cared about their families and treasured their leisure time.

Organizations and Attractors

Organizations can be similar to human personalities. They have their family histories and personal stories; decades later they may still be playing out the consequences of past traumas. I remember an old lady, quite comfortably off, who lived very parsimoniously, even chewing crusts of stale bread rather than throwing them away. She had lived through the Great Depression and World War II in England, one a time of crushing poverty, the other of rationing and deprivation. The traumatic memory of those events had never left her. In this area of her life she had become closed to change and trapped in the strange attractor of her past.

The French psychoanalyst Jacques Lacan noticed that people can even become trapped by a name. Suppose, as is sometimes the case, a baby is given the name of a dead relation. That child grows within a certain matrix of stories told about the dead relation: memories, anecdotes, and amusing or tragic stories. It is as if, when the child looks in the mirror, he sees not so much himself and his own face but a vague image of the person he is supposed to represent. Rather than identity being an interior matter, Lacan observed, in such cases the patient identifies with that exterior image—a dead relation or some sort of ideal that parents have projected onto their child. In other words, he did not feel himself as inhabiting his own body but as being elsewhere. He felt bound by certain dimly understood drives to fulfill a role and to become that person who can only lie outside himself. Again, what applies to an individual can apply to a business, an organization, or even a government.

It is often the case that a company calls in efficiency experts or business consultants to observe its operation and offer advice. Just how effective this proves to be depends on factors that are also seen in the relationship between psychotherapist and patient. Many therapists set great store by *the initial interview*—that first meeting between therapist and patient in which the patient attempts to position herself in relation to the therapist. Clearly this can be a very tense time. The patient is admitting that she is having problems in her life. She is asking for help and anticipates having to go into painful and embarrassing details. It is out of this tension that the therapist notices many of the patterns that have been underpinning the patient's past life. Does she relate to the therapist as an authoritative parental figure? Or as someone who can be seduced into giving in, making deals about fees, cutting corners, and arriving at compromises? Is she afraid that the therapist may not always be there for her at the same day and same hour? Does she feel that in some way she is being cheated out of the 50-minute therapeutic hour she has paid for? Will she adopt strategies to win a few minutes more? or attempt to invade the therapist's private life by discovering details of home, family, and background? Will the therapist end up colluding with the patient? or take a vicarious enjoyment in her shady sexual and business exploits?

Within that first therapeutic encounter the future course of therapy may be made or broken. It is as if the therapist is always in danger of being contaminated by what could perhaps be called the patient's attractor, that history of relationships and repetitive pattern of behavior. If the therapist is strong, firmly centered, experienced, and alert the therapy will go well. But sometimes a therapist becomes sucked into playing along with a lifelong survival strategy established by the patient. The patient may win over the therapist to her side, or lean on the therapist for months to come, or use the therapist to gain approval of her behavior with a partner or business figure.

The same thing applies to organizations. To the extent that they are gripped within their own history they are incapable of engaging fully in a creative act of growth and of maintaining flexibility in the face of change and uncertainty. Individuals and organizations that behave in repetitive ways are always following some limited set of goals and repeating their mistakes. They are similar to self-organized systems in the grip of an attractor. No matter if employees and directors come and go, no matter if computers are exchanged for typewriters, or even if the company moves from a Victorian building to a modern high-rise, a hidden magnetic attraction will still be present.

Some consultants refer to the "story" of an organization and the way this continues to play itself out decades later. A large organization operated with two corporate executive officers (CEOs) rather than the more common single CEO. Naturally this gave rise to all manner of tensions and conflicts within the organization that compromised its efficiency. It was no surprise to learn that, well over a century earlier, the company had been founded by two brothers at a time when their country was involved in civil war. It is as if some sort of memory was operating within the organization, a type of attractor that created dualism and division.

In such cases, feedback loops have become fixed and do not readjust to new circumstances. Likewise, iterations continue to flow throughout the system to support a set of fixed responses. Like a human heart that exhibits too much order in its rhythms, these systems have become overly rigid and no longer embrace the creative side of chaos. Maybe at some time in the past when the economic, business, or political environment changed, that organization closed itself off from the full potential of the outside world to the point where it now only engages the marketplace in a limited number of strategies.

On the other hand, as with any living organism, an organization may have a natural lifetime. Some wither and die. Others occupy a sort of fossilized position in the marketplace, like one of those curious animals found in odd ecological niches of the world. They may still be making money, yet generate little satisfaction for those who work within their walls. The organization simply "isn't going anywhere," and so workers become indifferent to its goals.

It is also true that an organization can undergo a radical form of renewal. It can grow creatively. It can accept the challenge of a changing world and employ the creativity of its employees to the full. But if such an organization wishes to adjust, learn, grow, and renew itself, it must be willing to go through a period of reorganization. This may mean opening up the feedback loops, changing the pathways whereby information and meaning circulate around the organization, maybe even changing the way computer systems, rooms, corridors, work hours, meeting rooms, and the like are structured. To carry out this renewal, the organization will have to face an initial period of chaos. Many people fear chaos because for them it means lack of control. Familiar routines may become disrupted. New relationships will have to be made. People may be required to learn new tasks, and a variety of formal and informal groups may have to be reorganized.

Lessons from chaos theory show that energy is always needed for reorganization. And for a new order to appear an organization must be willing to allow a measure of chaos to occur; chaos being that which no one can totally control. It means entering a zone where no one can predict the final outcome or be truly confident as to what will happen.

Yet, in the last analysis, all organizations and groups deal in human relationships. And this means they deal in fears, disappointments, and aspirations. It means taking into account those who, for several years, may have felt slighted, snubbed, not given proper respect, or not listened to. Change may offend vested interests and threaten those who simply want to keep doing the same old job. This is where the metaphor of chaos theory has its limits, for organizations are composed of human beings and not abstract sets of feedback loops. Human beings need to feel respected. Most of them like being part of a group. Most of them wish to feel challenged and their creativity fully used. What's more, human beings need to feel that there is a meaning and purpose to their lives. Part of this meaning comes from the warmth of their family and friends and part from their work. It is not money alone that attracts an employee or manager but the challenge of work. It is the possibility of learning new skills, of extending oneself, and of feeling that one is doing something useful and meaningful in the world. Highly creative people are willing to take a drop in salary to move into an organization or field in which they feel truly creative, or one that is ethically satisfying in that it does something positive for society or for the environment.

Many of the social and political movements that arose in the past decades spoke to people who felt themselves marginalized and disenfranchised—people of particular races or sexual orientations, women, or those who have particular mental or physical disabilities that prevent them functioning in the same way as the majority of the population. People may feel themselves discriminated against, often through the subtle and largely unconscious attitudes of others. It is only when we are open to change and renewal that we realize that we belong *inside* society, that a healthy and creative group, society, or organization is not something external to us but, in the last analysis, it becomes the expression of each one of us, and each of us shares in its meaning.

Action

If we can never have total certainty, and if our abilities to predict and control the world around us are inherently limited, then the metaphor of chaos theory will lead us to rethink what it means to take corrective action. What does it mean to make plans, execute policies, and aim at goals in a world that always contains a measure of uncertainty and ambiguity? In short, what guidance can chaos theory give us when we feel the need to take action?

Newspapers write of fighting crime, the war on drugs, the war on want, and now the war on terrorism. Doctors speak of taking aggressive measures in fighting a disease. Issues are to be *challenged* and *confronted.* The rhetoric of combat, conflict, and aggression is all around us and seems unnecessarily violent, considering that these are issues regarding social and medical matters. It suggests a mindset desperate to retain control over each and every situation, so that deviation from a preconceived plan, goal, or ideal is seen as involving something akin to a moral lapse that requires correction and punishment. Action of this nature cannot tolerate uncertainty. It uses the language of confrontation, a language in which problems are to be *dominated* and *overcome.* Such rhetoric is also used to whip up support at elections by suggesting that a wrong or inherent evil has been pinpointed and, like an enemy, is going to be beaten to its knees. This same rhetoric places issues and problems as lying outside us. It seeks to apportion blame to extraneous factors and is tailor-made for the creation of the "other"—ethnic, social, economic, or religious—group that can then be blamed for all of society's ills. Scapegoating has been going on for millennia. It is easier and more convenient to lump people together under a flag, skin color, or religion than it is to take into account the wide range of human individuality and diversity.

Once again we encounter a central issue of this book. It is that of objectifying the world and attempting to stand outside a system as a supposed omniscient observer. It is the action of distancing oneself and seeing the world in terms of "problems" and "solutions," instead of realizing that societies, cities, nations, and economic systems are immersed in complex webs of meaning that give them their cohesion and from which they take their values. People may be good or bad, stupid or creative, ignorant, uneducated, traumatized, or in some cases simply evil. We can never place ourselves outside the system as observers; our behavior, goals, and values are always set within that matrix of meaning that emerges out of the multilayering of family, group, society, global economics, and so on. Any policy or plan, any action taken, unfolds out of this matrix and its accompanying values and meanings. In turn it acts back upon it. Going to "the heart of the problem" may be important, yet it can also mean ignoring all the factors that gave rise to that situation in the first place, or to those factors that are ameliorating the situation at the present moment and causing it to persist. When we look at the world as object, or "problem," we forget that we too are an essential part of the pattern we see around us.

If we have an overly rigid approach to life we treat the world in a mechanical way. If a clock, or any other mechanical system, malfunctions we take it apart and look for the cause. Such a system is composed of parts connected together. When it doesn't work we suspect that one of those parts has failed or come loose. And so we take the mechanism apart and look for the bits that don't function.

This approach works perfectly with clocks, toys, car engines, and other mechanisms. But how well does it apply to a city, a society, a human being, a polluted lake, or the stock market? When we view the world as a machine, we think of it and act toward it in a mechanical way. When we deal with a machine we believe that every malfunction can be analyzed and reduced to a problem associated with some defect in a component. Such problems always have easy solutions because components can be repaired or replaced. And so we end up responding to the world in mechanical ways because we see it as no more than a particularly elaborate machine.

To build a clock or a car you take parts off the shelf and assemble them together. But in the case of self-organized and open systems the "parts" are expressions of the entire system. A river isn't composed of smooth water and vortices glued together. Rather, the vortex, while remaining stable and identifiable, is an aspect of the entire river. Likewise, the volunteer groups in a community are expressions of the cohesion and meaning of that town or city.

When a social or natural system malfunctions this can sometimes be traced to a fault in a particular aspect. More often it is a deficit of the entire system. Take, for example, the human body. Falling and breaking an arm or leg appears at first sight analogous to a defect within a machine—we can no longer walk, or lift things, because of a defective component. On the other hand this failure may be an expression of a long-term defect in the entire system. The leg may have broken because of osteoporosis—a lack of calcium in the bones. This could be the result of a faulty diet, but is most probably a calcium deficiency resulting from the body's metabolic changes caused by aging. Or a person could have fallen because he was preoccupied and did not notice where he was going, or because he had been drinking in an attempt to relieve a high level of stress. In turn, people's jobs and the need to make more money to support a particular lifestyle produce such stress. And so the failure of a particular component ends up being connected to many other factors and meanings.

When a group of people is exposed to the same virus some become quite ill, some experience a couple of days of tiredness and slight fever, and others notice nothing untoward. Why is this? Why do some people become ill and others remain well? Issues of the effectiveness of the immune system touch on a wide variety of factors: on the negative side, stress, lifestyle, and exposure to low levels of contaminants in the environment; on the positive, the ability to laugh, a life full of meaning, a deep interest in friends and relations, and a feeling of something positive to aim for in life. In a wider sense the health of the immune system becomes embedded in our work, family, and the values and structure of our entire society.

In turn, what applies to the human body and the life of the individual also applies to a society and an environment. When problems surface, the causes may be complex and interlocking. In so many cases they depend upon levels of meaning and contexts.

Chaos theory tells us that we may not always be able to "control" or "fix" a particular situation. We know that some systems are highly resistant to change. Others may be oversensitive so that a small interaction may flip the system in unpredictable ways. Rather than seeing such systems in mechanical terms it would be more effective to feel out and understand the ways such systems function at an organic level. We need to sense them as living, functioning systems, to see how they depend upon complex levels of meaning so that any action we take flows from an understanding of this underlying meaning.

Action need no longer be violent and confrontational. We don't need to think in terms of problems to be tackled, or of making war on defects. Rather we must work at many levels simultaneously—at both the practical level and the level of meaning, dealing with both particulars and generalities, looking at both a specific defect and the overall context in which this defect occurs. We must remember that whenever we look at some system outside ourselves we are also looking inward at ourselves, at our projections and prejudices and our fantasies of how things should be.

Psychotherapists know this when they say that the patient is there to cure the doctor! Our drive to correct and improve things must always be open to question. We must ask why we make such an effort to deal with the world. Are we reacting to environmental disaster out of fear and anger or out of a deep love and empathy with the natural world? Do we want to heal because we don't feel whole inside? Do we wish to improve the world around us because we feel inadequate? Do we engage in endless activity because our own lives are empty? Every action flows out of who we are and the meanings we value. We are constantly bringing ourselves to the world, and who we are and the values we hold are aspects of that world of which we are an essential part.

The move from certainty to uncertainty that characterized the twentieth century has brought with it a great responsibility. Each of us today realizes our connection to the society in which we live through countless feedback loops. Each of us helps to generate and sustain the meaning by which that society functions. What's more, chaos is no longer something to be afraid of; it is an expression of the deep richness that lies within the order of the cosmos and our very lives.

Seven

RE-ENVISIONING THE PLANET

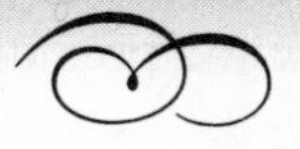

The human mind delights in creating all-embracing theories and definitive explanations. Yet, as we have seen in the preceding chapters, quantum indeterminism, chaos theory, the limits to language, and the incompleteness and uncertainty of mathematics all call into question the validity of such ambitious goals and plans. But here the reader could be excused for objecting that these case histories from science, philosophy, and mathematics, interesting as they may be, are remote from daily life. In most cases they are the end result of brainwork created by academics who work in ivory towers and look out at the world from a relatively privileged position. And, when we speak of a transformation in consciousness that began in the twentieth century, is this change confined only to an elected few, or does it apply to everyone?

This chapter discusses far more pressing issues—the global and local choices we face in our daily life, and decisions that will have an impact on our children and our children's children. These issues concern aspects of our daily living that our grandparents' generation took for granted but which we have now come to question.

The nineteenth century had been a time of vast horizons and empty spaces. Question marks could still be found on maps, and new lands were being opened to explorers and settlers. Over a century ago people believed that the earth and its resources were unlimited. There were always new materials to be developed and new energy resources to be exploited. There would always be something for everyone. Until the Industrial Revolution when machines acted to amplify human actions, a lifetime of human labor and effort had only a small impact on the earth. It was natural to believe that the human race would persist forever. Thinkers like Nietzsche and Bernard Shaw even believed that humanity was climbing onward and upward toward the age of the Superman.

All this changed during the twentieth century. The human race experienced the hubris of its earlier pride and arrogance. This change is symbolized by two remarkable images that have etched themselves deeply into our collective unconscious: a mushroom cloud and a blue ball in space. Both were the result of advances in science and technology. Both subverted our boast that humanity was capable of unlimited advance and progress.

The first, the mushroom cloud, stands for the atomic bomb and the generations of nuclear weapons that followed. For over half a century the world lived under the shadow of this cloud. During that period the *Bulletin of Atomic Scientists* set its symbolic nuclear clock on its masthead with the hands pointing at five minutes to midnight, indicating that the human race stood on the brink of a nuclear holocaust. Although a generation of American children was taught the nuclear drill of "duck and cover," scientists were soon pointing out the futility of the various emergency measures that had been set in place. The immediate effect of explosions and radiation was bad enough, but what came afterward would be far worse. As Soviet premier Nikita Khrushchev put it, after a nuclear war the living will envy the dead. Nuclear explosions would create vast dust clouds in the upper atmosphere so thick they would block out the sun's light and heat for years to come. A nuclear winter, a period of cold so profound and unremitting that it would wipe out not only the human race but also most life on earth, would follow.

The tension of the cold war is now behind us. But in a different form a nuclear threat still exists, not so much from the big superpowers but from smaller and less stable nations, and even organized criminal groups. Half a century of international tension has made us more aware of the fragility of life on the planet. Science has revealed other threats, from viruses to drug-resistant microorganisms. Recently a Swedish hospital discovered that hepatitis C had found a way of spreading to hospital patients not through the normal routes of intravenous injections of contaminated blood but as an airborne virus.

Ebola first emerged from the Ebola River region of Zaire in 1976. The death rate from the disease is 50 to 90%. There is no known treatment. AIDS is taking a terrible toll and its effect in Africa is proving to be as devastating as the Black Death that swept across medieval Europe. Yet the AIDS virus can only survive under optimum conditions. Imagine what would happen if such a virus could be transmitted by a flea or mosquito bite? Or if it were airborne, as was the case with hepatitis C in Sweden? Would that spell the end of our global civilization? Human life may be far more vulnerable than we imagine.

The second key image of the twentieth century, a photograph taken by American astronauts, is of planet Earth as a blue ball suspended in space. The fact that the earth is finite is something we all knew at an intellectual level, yet it required all the billions of dollars spent on the space race to remind us in a forceful way that we are all brothers and sisters. Native Americans say "all my relations," meaning humans, animals, fish, birds, insects, trees, plants, and rocks. That image from space reminded us all that we are inhabitants of a single earth and that its resources are not infinite. What is done in one place affects another. Smoke from the smelters in Sudbury, Northern Ontario, pollutes the northeastern United States. The rain that fell on my car in central Italy last night left a fine dusting of white mud—sand from the Sahara Desert carried by the wind.

When it comes to ecology and environmental pollution, there is no room for national politics. Wind does not acknowledge national boundaries, rain falls on international treaties. The destruction of the

Amazon rain forests is not an internal matter for the Brazilian government but an issue vital to the climate of the entire world. The choice of a family car or the act of switching on an air conditioner is no longer a matter of purely personal choice. It is on issues like these that the environmental movement takes its stand.

Environmentalism

In the mid nineteenth century only a small percentage of Americans lived in cities. Today the figure is around 50%. It is really only when people leave the countryside for a world of streets, high-rises, and shopping centers that they develop a nostalgic conception of an untouched world called "nature." For farmers and peasants, nature is something to contend with, something ever present. Nature feeds and nurtures, exhausts and threatens.

While it is true that William Wordsworth once wrote a sonnet deploring a railway planned to run though the Lake District of the northwest of England, country people do not generally romanticize their environment. To the citizens of our modern industrial world, however, nature represents an ideal, something "out there" that should remain forever pristine and uncontaminated. And, when we began to realize that that dream of lakes and woods, of birds and flowers was seriously at risk, the movement called environmentalism was born.

The word "environment" itself was not really used until the second half of the nineteenth century. It derived from earlier words like "environ" and "environing," terms reserved for notions of the surrounding and encompassing. The idea that an environment was a sort of entity, and that this environment could be at risk from human progress, only occurred to people in the mid twentieth century. Biologists understood the complex interlocking of natural systems, but the real interest in environmental issues dates from Rachel Carson's landmark book *Silent Spring* (1962), in which she pointed out the dangers of indiscriminate use of pesticides. Suddenly people realized that the idea of pollution did not apply simply to one lake or patch of woods, but to the entire environment. Ironically, the use of science and technology in our desire for ever more progress created an ecological threat even while it alerted us to the dangers it posed to the environment.

Rachel Carson's book appeared at a particularly appropriate time. It was at the start of the swinging sixties, that turbulent period when young people were questioning the wisdom of their parents, exploring alternative lifestyles, and discovering the power of political protest. Clearly environmentalism was an issue that politically aware people wanted to support.

As we look at our own contemporary environment, with its issues of global warming, genetically modified foods, the human genome project, and the depletion of the ozone layer we realize just how complex the world has become and how important it is to have clear and impartial information. We are now less certain about the consequences of that development we call "progress."

Most of us pay lip service to the idea that environmentalism is a good thing, that we should care for the planet, and that "someone" should do "something" about such issues as global warming, depletion of the ozone layer, and the destruction of forests and natural habitats. "What can we do about these important issues?" we ask. "Surely they can only be resolved through legislation and international agreements." We are certainly willing to support environmentalism during a dinner party discussion. But what else are we supposed to do?

On the other hand, each day we face a host of tiny decisions that call for a small degree of effort on our part. It's all too easy to throw an empty soup can into the garbage, or drop the newspaper we have finished reading into the nearest wastebasket. Yet we know we should take the time to sort bottles, newspapers, and cans into the correct recycling containers, and maybe doing so gives us a little thrill of pride that we are doing something positive for the planet. The problem is that we are never sure if what we are doing is important or not. In the long term, just how much impact will our individual actions have on the fate of the earth?

Shopping for the Environment

The problem is even worse the moment we step into the supermarket for the week's shopping. Aren't supermarkets, with all their packaging, creating waste? Are we using too much gasoline when we drive to them?

Suppose we have arrived and begin to shop. Things are not too bad at the meat counter, which now features a special section of meat that is free from hormones, and we believe that if the animals have been allowed free range the product will taste better. However, to produce the protein we get from meat, farmers need a great deal of available land to grow grass and hay needed to feed livestock. But crops grown on that same area for direct human consumption would produce far more protein. Does that mean we should all become vegetarians in order to avoid having to turn yet more forests and land into cattle farming areas? After all, we now realize that, with large areas of forest being converted into grazing plains, the lungs (the forests) of our planet are becoming compromised.

Now we reach for a box of eggs and become confused at the concept of "barn eggs," "free range," or even "organic." Just what do these terms mean in practice? Are these more expensive eggs any better for us? Does spending a few cents more on them have a beneficial effect on the planet? Or is the whole thing no more than advertising hype designed to seduce shoppers into parting with more money in the belief that they are buying a more healthful product?

If we have a baby in the house we reach for a packet of disposable diapers. These are a truly great invention because the semipermeable barrier means that the baby's bottom stays dry while the diaper itself absorbs urine, so no more diaper rash. On the other hand, millions upon millions of disposable diapers are destined to be buried in landfills where they take hundreds of years to degrade. So why not go back to using cloth diapers and washing them out? But boiling the water, washing with detergent, and adding effluence to the sewage system puts another sort of strain on the environment.

And so we go from aisle to aisle, shelf to shelf, facing a host of tiny decisions without ever fully understanding the implications. And then comes the last straw. We stand at the checkout and are asked, "Plastic or paper?" What on earth are we supposed to answer? We've been told that plastic is a "bad thing" and so we'd better choose paper bags. But paper bags often get thrown away. Their manufacture is a polluting process and their production involves cutting down trees. Plastic bags, on the other hand, are reusable and once they have ripped they can be thrown away as biodegradable matter. On the other hand, their manufacture involves the consumption of the world's nonrenewable oil resources. The supermarket has put the onus of choice on us, the consumer, but in the end the best we can do is toss a coin and choose one or the other. Of course people who live in smaller communities still make their purchases at local shops where they can use cloth bags or baskets to carry home their daily groceries.

There is a host of similar questions that arise, and in each case we want to know what the right action is for each situation. Even experts are divided on many of these questions. The photographer and environmentalist Mark Edwards has spent his life documenting what has been happening to our planet, as well as noting what has been occurring within the various environmental movements. For Mark these issues are not so much questions as "disturbances." They are part of the many small issues in daily life that worry and concern us—in other words, that disturb us.

The problem is that we have been so accustomed to living with certainty that we assume every crossroads must contain a right and wrong road. In the bygone world of Aristotelian logic if one choice is right then the other must be wrong. But in our modern world neither choice may be exactly right. The issues have become so complex that every action resonates through the environment in unpredictable ways. When it comes to costs and benefits, it is increasingly difficult to put an exact price on the consequences of our action. What is more, there may be no "correct" fixed answer. Take, for example, the issue of wastepaper. Areas of the world cut down trees to turn them into pulp for our books, newspapers, packaging, and a host of other products. In turn these areas are replanted, often with monocultures. Naturally people turned to the idea that paper should be recycled and anyone who was ecologically minded made sure that the cartons, books, and paper supplies they bought were all made from recycled paper. But now some experts are suggesting that, in the ecological accounting book, it makes more sense to burn wastepaper to generate heat for industry or a city central heating plant than to recycle it. And so one answer that was the "right" one is suddenly replaced by another. Is the first answer, recycling, then "wrong"? Did we make a mistake? Or will expert opinion change yet again? Is this another disturbance we must put up with?

Experts

In Chapter 5 we asked if someone, an art critic or gallery director, would provide us with a set of standards as to what is good and bad in art. No answer was forthcoming, other than our own obligation to inform ourselves and make our own judgments. But now consider something much more serious. Areas of the planet are under environmental threat. We want to do the correct thing. But how are we to know what we should do? The world has grown so very complex. Who is to advise us? Where can we find an impartial expert?

There is a story that after Mahatma Gandhi was shot, the new prime minister, Pandit Motilal Nehru, asked an Indian sage what he should do. "Right action," came the reply. The same advice could well be given to our present leaders, society, and ourselves. In the face of environmental dangers, decay of the inner cities, and international tensions what we need is *right action.* But, in a practical world, and faced with a number of alternatives, just what is the right action? In a world that has become so incredibly complex how can we be sure of the implications of our daily decisions? What are we to do? These are our contemporary issues of uncertainty and tension or, as Mark Edwards puts it, of disturbance.

Even at a fairly primitive level our minds cannot tolerate uncertainty. When carrying out a task such as solving a problem in algebra or getting a photocopier to work our brains don't like to be stumped or put in the position of "what should I do next?" Such issues create a sensation of tension and discomfort. Experiments by psychologists indicate that when we reach a point of uncertainty in carrying out a task we tend to patch over it unconsciously by inventing an arbitrary rule.

A small child may appear to be constantly making mistakes in arithmetic, but closer investigation will probably show that she is ignorant of a particular mathematical step and so she has invented a rule—albeit incorrect—which she then uses consistently in her calculations. Rather than stopping and having to support an inner tension, the mind patches over things and keeps running.

Likewise, when we are faced with the disturbances of daily decisions, of weighing up alternatives and wondering which is the correct ethical choice, we prefer to pass the buck rather than tackle the issues ourselves; and so the answer comes: "Put the responsibility on someone else's shoulders." "Ask the expert." "Our tax dollars support the government so why don't politicians pay some really bright scientists to come up with the right answer and tell us what to do?" "After all, scientists split the atom and put men on the moon, why can't they tell us which sort of bag to select at the supermarket checkout?"

But where are we going to find these experts, these dispensers of ecological wisdom? When the American colonies decided to "dissolve the political bands" that connected them to the British Crown they were careful to do so in a spirit of *right action.* The founding fathers of the United States of America set down a carefully worded argument as to why foreign authority could not be justified. This Declaration of Independence drew upon the best minds of its day, with Thomas Jefferson as its principal author.

Law, to Jefferson, was not simply a matter of torts and contracts but a way of understanding human culture, history, values, and meanings. In writing this declaration, Jefferson drew on this philosophy. He also had the help of another exceptional man, Benjamin Franklin, who added that marvelous phrase, "We hold these truths to be self-evident," when referring to the rights of life, liberty, and the pursuit of happiness.

At a critical time in its history, a new nation could turn to leaders and philosophers concerned with truth rather than power, fame, and popularity. The founding fathers were not bothered about placating lobbyists and vested interests or worried about donations to a political party. Their main interest did not lie in pleasing the public in order to be reelected for another term in office. Rather they sought to make wise and impartial decisions and looked to a future that was wider than the next election.

Democracies call upon us all to make informed decisions. Whether it is a matter of electing a politician, voting in a referendum, or picking the right bag for our groceries, we are obliged to think wisely and assess the information before us. Yet the issues of today, such as globalization, the economic disparity between first and third worlds, global warming, depletion of the ozone layer, decay of the inner cities, and drug dependency, are far more complex than those that faced the founding fathers of the United States. To whom are we to turn for our answers? From where can we obtain clear and unbiased information? How are we to access the pros and cons of genetic engineering, nuclear power, or trickle-down economics?

Ours is a period when wisdom, judgment, honesty, and unbiased information—in other words, certainty—is badly needed. We expect to be informed by newspapers, radio, and television. We need information that is presented clearly, with mature reflections and informed comment. Yet all too often we are offered the soothing words of the professional "expert." When a news story breaks—a stock market crash, air disaster, nuclear accident, disease epidemic—an expert is always on hand to deliver opinions in palatable sound bites. While there are many highly professional and educated television journalists and producers, television news is subject to a major constraint: it must do well in the ratings. Such a format is not easily designed for scientists or academics who wish to qualify their opinion with "maybes," "possiblys," "on the balance of probabilities," "in certain cases," or "in this particular context." Far better to present an issue as controversy. By reducing things to black and white an issue can be dramatized with two "experts" who battle it out for a minute or two in front of a smiling moderator.

Newspapers have time to be more reflective, yet even they have their advertisers, as well as the political agenda of the proprietor to think of. Readers of the major newspapers are the victims of circulation wars, so that sometimes the real gems of writing and reporting can only be found in small-town newspapers that do not have to compete at the national level.

The Political Line

From time to time during the twentieth century quite ordinary people have been swept up in a degree of paranoia. "Is the government lying to us?" people have asked. "Who was really behind the assassination of President Kennedy?" "Is it true that the authorities are flooding the ghettos with drugs to keep poor people complacent?" "Is the government keeping quiet about UFOs?"

We may laugh when we hear stories about government cover-ups involving the abduction of, and experimentation on, ordinary citizens by aliens. On the other hand there have been several unpleasant cases of official misinformation and misdirection. For example, take the outbreak of mad cow disease (bovine spongiform encephalitis) in Britain. This disorder was first recognized in the 1980s when outbreaks were identified in several parts of the world. It is transmitted through cattle feed that contains ground-up animal parts. After the disease had been identified, scientists became concerned that it could be transmitted to humans who ate the meat from infected animals. There was also evidence that the disease was present in other species, such as sheep, so that it wasn't sufficient to refrain from eating a hamburger or a steak; other foodstuffs could contaminate one as well. Even vegetarians and health food consumers were not free from risk, for many confectioneries and other substances contain animal by-products, such as the gelatin used in candies, jellies, vitamins and drug capsules, and so on.

Mad cow disease was bad enough in the United Kingdom, but the real scandal was the reassuring voice of the British government during the early years of the outbreak. Measures were in hand, everything was under control, and British beef was safe to eat, people were told. There was no risk to humans. The British people should not listen to unfounded rumors. And so when the European Community placed a ban on British beef there were massive demonstrations by British farmers and threats of boycotts against European products.

At the end of 2000 the results of a long enquiry were published and a former prime minister, John Major, was forced to make a public apology. During that same period, it later transpired, the British government had known about the spread of the disease and the risk to human beings; nevertheless, it had continued to issue bland reassurances and had even attempted to discredit those scientists who urged caution. Now people in Britain have begun to die from the effects of the disease, and since the effects are slow acting, no one really knows how high the mortality figures will climb.

Another bitter irony arises when we recall the peace conference at The Hague in 1899 and its desire to outlaw certain weapons of war. One hundred years later, newspapers were filled with stories of illness and death caused by the use of depleted uranium (DU) in projectiles used by North Atlantic Treaty Organization (NATO) troops in Kosovo and Bosnia. A NATO memo, issued in July 2000 reads, "It is clear that the medical hazard from DU is negligible." Another NATO information package "Medical Implications of Depleted Uranium" concludes that risks are "Overall negligible" and states "No further action recommended."

It is certainly true that picking up a lump of depleted uranium poses little in the way of a health hazard. What those reports did not mention was that when one of the 31,500 projectiles used in Kosovo or the 10,800 employed in Bosnia hit its target the uranium was vaporized and the subsequent inhalation or ingestion of uranium dust or particles can give rise to leukemia; cancers of the lungs, kidneys, and thyroid; genetic anomalies; and the general depletion of the immune system.

The World Service of the British Broadcasting Corporation also quoted Dr. Asaf Durakovic, a professor of medicine and former U.S. Army colonel, as finding a "significant presence" of DU in two-thirds of the Gulf War veterans he had examined. In some cases particles of uranium were still lodged in their lungs, where they pose a constant risk of cancer.

Ivory Towers

If the statements and promises of politicians and television experts are suspect, must we then turn to the traditional repositories of wisdom in our society—the universities? Visit the University of Virginia at Charlottesville, designed by Thomas Jefferson, and notice the way the architecture complements its function. Universities are enclosed orders, a group of buildings clustered around a central grassy courtyard. They were built as places where people could gather together to reflect on knowledge. Leaving the noise and bustle of the town outside they became islands of tranquility that prized the excellence of scholarship. A variety of different subjects were taught in the colleges. These were places where people could study, live, and eat together. Scientists, philosophers, theologians, and those in the humanities could meet daily at dinner and consider, debate, and reflect on issues of knowledge, morals, ethics, and the condition of society. As the economist Arthur Cordell puts it, universities were the flywheels that damped out the more eccentric oscillations of society. Universities were places where new ideas could be tested. They were free from censorship and bias. They were the debating society for national and global issues. If the uncertainties of ecology were to be resolved, then this would occur within the university. That was the grand vision. That's how things were supposed to be before dreams came crashing down around us and glass and concrete knowledge factories were built.

The first universities grew out of early medieval *studia generalia* and developed into corporations of students and masters, with groups of lodgings for scholars from all over Europe. In 1135, for example, the philosopher Peter Abelard moved to Monte Sainte Genevieve. Soon some of the greatest minds in Europe came to settle on the left bank of the river Seine to hear him lecture. The University of Paris was born out of this informal congregation of scholars' lodgings. Such early universities were given charters by kings or popes to make them totally self-governing and free from the laws of the town and country. Scholars were free to teach and to question. The great universities were a free and independent haven for learning, and acted as repositories of knowledge.

Today education, in the United States at least, has become big business. Students are viewed as clients who demand efficiency and cost-effectiveness from their courses. A university degree is supposed to open the door to a better-paying job by emphasizing skills and abilities related to jobs and professions. Universities compete for the student market and are managed like any other corporation in terms of efficiency, cost cutting, and profit making. And so, while innocent academics were pondering the physical and moral questions of the age, the institutions changed around them to the point where these same academics were no longer in control.

Today large corporations endow professorial chairs, and universities are dependent on donations from businesses and influential individuals. Major research grants are funded directly, or indirectly, by the military. In a highly competitive world lecturers are desperate to gain tenure, and that means pleasing the students who grade them as teachers, while at the same time showing to the university authorities that they do not intend to make waves. In these and so many other ways the universities have become dangerously compromised.

At the same time individual scientists, economists, and medical experts who are called on to make judgments and disseminate information are increasingly mistrusted by the general public. When experts are quoted in a newspaper or interviewed on television, we wonder just where their vested interests lie. Who awards their grants? What offers to serve on corporate boards have they had? Imperceptibly the academic world changed and universities today are no longer totally trusted as places of free and open debate.

False Memories

Maybe the academic world has never been ideal and Byzantine plots and jealousies have always lurked within its ivory towers. During the McCarthy era, for example, academics were willing to make public statements in order to distance themselves from their colleagues who were politically suspect. The physicist David Bohm, however, refused to give names of colleagues when questioned by the House Un-American Activities Committee. In my book, *Infinite Potential: The Life and Times of David Bohm* (Reading, Mass.: Addison Wesley, 1997) I relate an account of a subsequent seminar held at Princeton University, essentially to discredit some of Bohm's work in physics. While it was to be a debate on technical matters in science, reputable scientists could not help using such terms as "Trotskyite," "traitor," and "fellow traveler." At the end of the meeting J. Robert Oppenheimer went so far as to proclaim, "if we cannot disprove Bohm, we must agree to ignore him." In part Oppenheimer was objecting to Bohm's scientific ideas but also, in part, to association with a politically "tainted" figure.

A more contemporary example is the debate on false memory syndrome. Child abuse, both sexual and physical, can do enormous damage. Those who have been raped or assaulted in early childhood often repress the memory, which later emerges as psychophysical symptoms. It was Freud who first alerted us to this phenomenon and pointed out that psychoanalysis may aid in resolving such painful issues. However, during the 1980s, the phenomenon of repressed childhood rape became over-fashionable. Schools of therapy, including hypnosis, claimed to allow patients to move back into early infancy and discover examples of sexual abuse by friends, relations, and even parents. There were even cases when children said they had been made part of satanic rites, or when sexual abuse had involved entire nursery schools.

Some researchers became suspicious of the more elaborate stories and began to look into the way these accounts had been obtained. They found that, in some cases, a patient, placed in a vulnerable position, would look to the therapist for subtle clues as to how to proceed and what to say next. If the therapist happened to be a proponent of theories of parental sexual abuse, then sure enough the patient would start to "remember" details of such abuse that never really happened and weave a consistent story. Similar "false memories" can also be generated when a hypnotist or therapist urges a patient to remember details of a serious traffic accident, robbery, or act of violence.

During the 1990s researchers attempted to bring this problem to light through open debate. The University of Montreal, for example, set up a meeting to discuss false memory syndrome. The meeting was a disaster. Some academics bussed in supporters who shouted down several of the speakers. Verbal abuse was exchanged and the possibility of any open debate abandoned. It was clear that where some issues are concerned the universities are no longer havens for free and open discussion. Indeed, some therapists and academics are afraid to publish research in certain areas because of the possibility of personal abuse or attacks.

In September 2000 I was responsible for putting together a roundtable meeting of academics from a variety of countries and disciplines to reflect on the future of knowledge, education, and the universities. The general conclusion we reached was that, in the main, universities no longer fulfill their role as centers where experts from a wide variety of fields can debate and discuss ideas together. The external pressures placed on the universities have compromised their ability to do free and high-quality research. This is particularly true of the so-called orchid disciplines; those subjects and areas that do not guarantee immediate and practical return.

The participants felt that academics should have an unwritten contract with the future—to provide an education that will open young people's horizons, educate, and inform them, and so produce more rounded individuals.

What can replace the universities? Maybe smaller formal and informal academies[1] where people could gather together and debate, but these too could have their drawbacks, for they will always be in danger of falling into the same traps as the universities. When we see even the

---

[1]Academies are often loosely knit, informal groups where people can come together to exchange ideas, stimulate, and challenge each other. The supreme example is the Platonic Academy that flowered in Renaissance Florence under Marsilio Ficino and the patronage of Lorenzo di Medici. It was located in a villa in the hills above the city where philosophers, poets, and artists (of the caliber of Michelangelo) could meet in a collegial way. The same function was served by the salons of nineteenth century France.

Black Mountain College in North Carolina was a catalytic center of exchange in art, music, and literature during the early 1950s. It was there that the poets Robert Creeley and Charles Olson experimented with a freer, although disciplined, approach to poetry, and through the *Black Mountain Review* gathered together those of similar minds such as Alan Ginsberg, William Carlos Williams, and Gary Snyder. The Black Mountain center was also where the composer John Cage met the artist Robert Rauschenberg. The two were to exert a strong mutual influence on their respective arts.

The author is at present attempting to create an academy in the small village of Pari, Italy, where he lives. The Accademia dei Pari is a group of artists, psychologists, and academics who meet from time to time to discuss our contemporary society and its values.

universities failing us, the uncertainties of the future make us feel even more isolated.

Risk Analysis

It is particularly ironic that the decline in wisdom and conviction and the rise of "opinion" and the "expert's" sound byte, should occur just at the time when we most need guidance. This is a period in earth's history when human beings are making enormous and sometimes irreversible impacts on the planet. How do we estimate the damage? How do we draw up the balance sheet of costs, dangers, and ecological disasters? How can we gain an objective and scientific assessment of risks and their implications?

Mathematicians and engineers have created a branch of science known as "risk analysis," which enables them, for example, to calculate the probability of a nuclear accident occurring within a certain time period, or to determine how safe it is to travel by car, train, or aircraft.

Suppose engineers have designed a pump that carries cooling water to a nuclear reactor. From the many tests done on prototypes they have a pretty good estimate of the effective life of the pump and the probability that it will fail within, say, the next 12 months. To be on the safe side they install a second pump that operates independently of the first. There is a small chance that one pump may fail on a particular day, but the chance of both failing at exactly the same time is remote. Nevertheless, engineers are able to calculate the very tiny risk of a simultaneous pump failure.

If a pump fails, a warning light alerts the operator, or activates a computer system, to take emergency measures. Engineers must therefore calculate the probability that two pumps fail simultaneously plus the probability that, despite regular safety checks, the alarm system malfunctions at the same time. In this way they are able to calculate the chances of every conceivable combination of failures, at the same time always making sure that independent back-up systems exist. They also work out a variety of "worst possible" scenarios—what if a jet plane crashes into a nuclear power station?—and then design fail-safe systems that will kick into action should such a potential disaster occur. The final result of their calculations tells us the, albeit remote, risk of a serious malfunction in a nuclear power station.

Risk analysis is applied to a variety of situations—loss of control in a jumbo jet, the possibility of two trains colliding in a metro system, the escape of deadly viruses from a research laboratory, the accidental release of genetically modified substances into the environment, and so on.

Because risk analysis involves a great deal of analysis and calculation and its end result is a series of numbers indicating risk, the approach appears "scientific" and lulls us into feeling relaxed about things. Finally, we feel, science and reason have placed a fence of certainty around chance and hazard. But we should never forget that there will always exist two important areas of uncertainty. The first is that the approach on which risk analysis is based is that of anticipating all possible failures and accidents. In practice this means everything that an engineer can imagine will ever go wrong. What can't be accounted for are missing factors and things overlooked.[2]

The other, and more serious, uncertainty is that low-risk systems only operate properly in the context of a well-run and well-funded infrastructure. Provided that an airline company, or the owners of a nuclear power station, are highly reputable and have no serious cash-flow problems, things go smoothly. But what happens when institu-

---

[2]There is a tragic irony that these sentences were written before the terrorist attack on September 11, 2001, that destroyed the twin towers of the World Trade Center in New York City. Risk analysis may be good at making a quantitative estimate of the probability of an anticipated event (however remote) but will never be able to "predict the unpredictable." Certainly the possibility of fire in the building or even damage caused by the collision of a light aircraft had been taken into account by the architects who designed the building. Its steel frame was protected by cladding able to withstand high temperatures over a certain period of time without the steel losing its strength. What had not been anticipated were the consequences of an impact by a larger airliner filled with fuel. Indeed, according to a report in the British newspaper *The Guardian*, a spokesman for the U.S. Insurance Information Institute said that "the possibility of the loss of both structures was seen as so remote that cover was not taken out on those lines." In a previous year a ship as vast and well designed as the *Titanic*, with its waterproof bulkheads in case of damage below the water line, was considered "unsinkable."

tions come under financial or political pressure, when they engage in cutthroat competition, when they operate in an environment where bribes and corruption are the rule, or when operators are overworked and poorly paid? These are the conditions under which bad mistakes can occur. In every situation, and no matter how many automatic fail-safe systems are installed, the human factor can never be ignored, and it remains unpredictable.

Take, for example, air traffic controllers, those invisible seat belts we all rely on when we travel by plane. Effective air traffic control is vitally important to avoid collisions during approaches and takeoffs near airports. Yet, as I write this book, controllers responsible for the airspace over London are complaining of overwork and high stress levels. They predict that a serious air collision will occur unless their working conditions are improved. In this case the risk analysis has been carried out and various technical components are all in place and working perfectly, but the human operator has become the weak link in the chain.

The Unforeseen

This chapter follows our twists and turns as we seek certainty and reassurance in the face of a variety of ecological crises. Universities and experts, we discover, have become compromised. The issue is not so much that any individual expert could be in someone's pocket, but rather that ordinary people are no longer so ready to trust an expert's advice. We saw that risk analysis, while producing scientific-looking results, can sometimes founder because of missing information or that ever present human factor. So can anyone reassure us with a definitive answer?

As we saw in the preceding chapters, twentieth century science caused us to confront issues of uncertainty and limits to our ability to predict and control the world around us. The natural world sets a barrier on how much we can know and how accurately we can anticipate the future. In case after case our attempts at intervention have been subverted. But we really did not need a couple of generations of theoretical physicists to tell us that our best-laid plans can sometime be sent head over heels. Common sense warns us that rational approaches such as risk analysis will always have their limits.

Our future contains a number of important issues: global warming, human cloning, depletion of energy sources, sustainable economics, overpopulation, worldwide distribution of food, depletion of the ozone layer, human–computer interfaces, and so on. We need to make wise decisions. But how are we to act in the face of uncertainty?

We are told that genetic engineering will increase food yields and help us to eliminate certain medical disorders. We are told that once the technical problems involving nuclear fusion have been solved there will be abundant energy for the entire world. We are told that human capacities will increase once individuals are linked directly into a computer; that ever newer materials will be produced for our homes and automobiles; that new communication technologies will revolutionize the way we work and learn. We are told that human beings will colonize space and one day may travel to the stars. We are told that there is no limit to human ingenuity, no limit to intellectual advancement and technological progress.

On the one hand the future holds a variety of promises, on the other, a number of threats. Just how are we to view that future? What position should we take? Where should we stand?

Often the best position from which to view the future is from the past. Let us look at the history of a series of advances that promised to improve all our lives—Freon, leaded gasoline, DDT, and antibiotics.

Freon

The coils inside a refrigerator need a gas that will both liquefy and evaporate easily so that heat can be extracted from the food and radiated from the coils in the back of the refrigerator. Ideally, in case of leaks, it should be safe for humans and chemically inert. Such a gas was discovered in the 1930s by a chemist named Thomas Midgley. Trademarked as Freon, it consists of what are known as chlorofluorocarbons. It was a dream substance used not only in refrigerators, but also in air conditioners and as the propellant in aerosol cans.

At the time Freon appeared perfectly harmless. It didn't corrode, and if some of the gas did escape it had no odor to contaminate food, while its chemical inertness did not present a fire or health hazard. It was only decades later that chemists began to suspect that chlorofluorocarbons might affect the ozone layer. As a result of the sun's action on the upper atmosphere, three oxygen atoms combine to form a single molecule of ozone. Ozone is of critical importance to life on earth because it acts as a sunshade over the earth and filters out the sun's harmful ultraviolet rays. But sunlight also breaks down Freon molecules in the upper atmosphere and releases chlorine, which then attacks ozone molecules.

It was only with the advent of space probes that scientists discovered how disastrous Freon's effects had been over the last decades. All the Freon from aerosol cans, discarded refrigerators, and air conditioners had collected in the upper atmosphere where it was attacking ozone. As a result, it had created a large hole in that part of the ozone layer that lies over the South Pole, and a smaller hole over the North Pole. If this depletion had continued unchecked—and ozone is still under attack from residual chlorofluorocarbons in the upper atmosphere—it would have endangered all life on earth, increasing the rate of skin cancers and promoting genetic damage.

Lead

In the first chapter we listed the automobile as one of the key factors in shaping our modern world. The availability of cheap and rapid transport transformed society and enabled people to work in cities and commute to suburbs. Indeed, the suburb itself, along with the highway and drive-in were all spin-offs of the automobile. What's more, the demand for ever more gasoline has transformed global politics by increasing the strategic importance of oil-producing nations. The automobile is also the major cause of death among young people. More Americans have been killed in automobile accidents than have died in all the wars the United States has fought since independence.

The impact of the automobile on society demands a book in itself. But just take one related issue, the attempt to make engines more efficient and powerful. As automobile engines became bigger, the problem of "knocking," or pre-ignition of the gasoline/air mixture in the engine cylinders, became a major problem. For the engine to work efficiently, this mixture had to explode at just the right moment—when it is ignited by the spark plug—so as to force down the piston and turn the engine. But as an engine heats up, this mixture can also explode spontaneously producing a "knock" or backfire that reduces efficiency.

The solution to the problem of knocking was discovered in 1921 by Thomas Midgley who, not content with discovering Freon, also found that when a compound of lead is added to gasoline it prevents pre-ignition. Thanks to Midgley's discovery automobiles could become bigger, faster, and more powerful. Two generations of motorists worldwide would have thanked Midgley had they but known his name.

Then, in the 1970s, scientists became concerned about the pollution of the atmosphere from automobiles. Cities had banned the use of smoke-producing fuels for heating, and the age of Sherlock Holmes's London fogs was past. But on hot days people now suffered from the choking haze produced by car exhausts. One of these effluents was lead from gasoline, and experts wondered about the effects of prolonged exposure to low lead levels, particularly on young children. Some scientists now believe that lead pollution is responsible for the decline in IQs of American children born between 1950 and the 1980s (when leaded gas began to be phased out for automobile use).

Having discovered both Freon and leaded gas, it is no wonder that Midgley is described by the author J. R. McNeil as having had "more impact on the atmosphere than any other single organism in earth's history."[3]

DDT

DDT was first synthesized during chemical research at the end of the nineteenth century, but its powerful effects as an insecticide were not

---

[3]J. Robert McNeil. *Something New Under the Sun: An Environmental History of the Twentieth-Century World* (New York: Norton, 2001).

discovered until 1939. At that time DDT held the promise to eradicate the carriers of malaria, plague, typhus, and yellow fever, as well as protecting crops from insect destruction. At one stroke the world was given the chance to eradicate several deadly diseases and increase food yields to the hungry.

And what about side effects? Insects are incredibly tiny; therefore what is lethal to a mosquito should have no effect on a human being. And so the use of DDT appeared perfectly safe, with the result that it began to be used indiscriminately. It was only after more careful research that biologists began to suspect that all was not well. To begin with, insects developed a resistance to the chemical, which meant that the doses of pesticide had to be doubled or tripled. Then a more serious issue emerged. Insects are tiny but they form the diet of fish and birds. When an enormous number of insects are eaten each day, DDT residues accumulate in the food chain until they compromise the health of higher animals—a famous example was of eggs with shells so fragile that they fractured when the birds attempted to sit on them.

The result was that the predators that feed on insects began to die off. Then, because insects breed much faster than birds, within a few generations pesticide-resistant populations of insects bounced back with a vengeance to occupy an environment lacking in natural predators.

Antibiotics

During the first decades of the twentieth century every hospital had a septic ward for patients in danger of dying from septicemia. By mid century, such wards had ceased to exist. This medical miracle was due to penicillin, the first of the antibiotics, developed in commercial form during World War II. Penicillin revolutionized the treatment of infectious diseases to the point that what had previously been life threatening could easily be cured at home with a course of antibiotics. Inevitably doctors and hospitals began to rely upon antibiotics as a panacea for infection.

On the face of it antibiotics are one of the great triumphs of modern medicine. How on earth could there be anything negative to say about them? It is only relatively recently that the medical profession has become concerned with the indiscriminate way antibiotics have been used, and fear that yet again humans have been too enthusiastic in putting all their eggs in one basket.[4]

As with any form of life, disease-producing organisms are not all genetically identical. While an antibiotic will wipe out most of a population of bacteria, a few hardier strains may survive. As time goes on, these bacteria begin to multiply to the point where a new drug-resistant strain dominates. This evolution of drug-resistant strains is also encouraged by those who never bother to take their full course of treatment—as soon as they feel a little better they throw away the bottle. Drug-resistant diseases are also prevalent among drug addicts and street people whose lifestyle leaves them open to a large number of opportunistic infections and who, in turn, do not seek proper treatment.

At this point, chemists race to develop a variant of the antibiotic that is lethal to the new resistant strain. But the war between science and microorganisms cannot continue indefinitely. Already tuberculosis, a disease that had more or less been eradicated from the industrial world, is reappearing in a strongly drug-resistant form. Ironically hospitals themselves, which once relied on the widespread use of antibiotics, have become potentially hazardous places in which to be ill. Statistics from Europe suggest, for example, that in normal pregnancies it is safer to give birth at home than in a hospital environment.

Of Mice and Men

Despite our inventiveness and the sum total of our scientific knowledge, the control we assumed we had over the world around us, and our ability to plan for and anticipate events is less secure than we sus-

---

[4]Doctors even prescribe antibiotics for viral infections while knowing perfectly well that antibiotics have no effect on viruses! If criticized they would probably argue that they are concerned about opportunistic secondary infections, yet would have to admit that they have no evidence that their patient actually has such secondary infections—it would be "just a precaution."

pected. New technologies arise, new scientific breakthroughs occur. Yet for every benefit we reap, there may also be unforeseen risks or side effects no one has anticipated.

Is the real issue the particular chemical, additive, pesticide, or antibiotic? Or is it their thoughtless and indiscriminate use? Is technology our enemy or is the enemy uncontrolled human thought?

If we cannot predict the future at least we can be sure about certain trends and alert ourselves to difficulties that lie ahead. As we face them we will be far less confident in our powers.

This, I believe, can also be a positive step. We must heed the warnings that our knowledge is not supreme. We must realize that we cannot sweep away problems that face us with another dose of technology. It is true that technology will always be essential in our modern world, but it must be kept in its proper place. We must treat it with respect and use it with wisdom. We must acknowledge our limitations and proceed with caution. We are like people lost in a dark wood who must move ahead in a tentative way, looking for pitfalls, groping around in the space before us, and ensuring that our footing is always safe.

In prehistoric times the world's population stood at a few million. Then, with the coming of agriculture around 1000 B.C., it increased dramatically to 150 million. This population continued to grow, but only slowly. In the seventeenth century, for example, it took 200 years to double the world's population, and the 1 billion mark was only reached in 1825. Today the world's population stands at 6 billion.

Thanks to the miracles of modern medicine, infant mortality has dropped dramatically. In Western Europe there are 6 infant deaths per thousand births, while in Africa the number of deaths remains around 90. The increasing number of children reaching maturity and the eradication of a number of infectious diseases and of epidemics have resulted in an explosion in the size of the world's population. If present trends continue this population will double in less than 50 years.

In the face of such a population explosion, food and energy become critical issues. Some experts feel that the only way to feed the world is to rely on ever more intensive farming methods. This means using pesticides, herbicides, fungicides, and genetically modified crops. Rather than farmers growing several varieties of a fruit or vegetable, scientists are designing crops that will meet optimum criteria, regarding not only their nutrient content but also their ease of harvesting and packing, shelf life, resistance to spoilage, uniformity of size, and so on. In more and more cases only one variety of seed is used. But this makes the entire crop more vulnerable to disease and insect damage. In addition, these new crops require the use of pesticides and herbicides, as well as fertilizers and ripeners. As a result modern farming methods are making much of the third world dependent on the products of the chemical industry. In turn, this requires a change in traditional farming methods, an alteration of entire social structures and reliance on the products and assistance of the industrial nations.

When it comes to animals, intensive farming, with large numbers of animals kept in close confinement, encourages the rapid spread of diseases and therefore requires the use of antibiotics. Hormones also speed up weight-gain and increase milk production. In addition, intensive farming produces environmental damage through soil erosion and pollution of lakes and rivers.

Both complexity theory and common sense tell us that diversity is the key to survival in the natural world. Yet more and more today we seem to be forced by circumstances to put all our eggs in single baskets. This can only spell disaster down the line. Natural systems have a built-in redundancy. If one part fails, others can take over. Block an artery in the human body and blood will continue to flow through smaller vessels that rapidly adapt to the increased blood flow. Plant several varieties of potatoes in a field and when a fungus or a virus strikes down one variety at least the others will survive. But when there is only a single system in the game then any failure or difficulty could be catastrophic.

Take as a particular example the hard disc on your computer. While you are writing an essay or an email message, information is being stored at particular locations (addresses) on this hard disc. When you then reread what you have written, the computer jumps to those addresses and displays the information on the screen. However, when local damage occurs on your disc, as sometimes happens, some of these addresses become unreadable and part of your work is lost—the hard disc has no built-in redundancy. Contrast this with the human brain. The words of a song, the memory of a face, and the tacit knowledge of how to ride a bicycle are not stored in particular groups of cells, but appear to be distributed over the brain. After a head injury a person's abilities may be somewhat compromised; on the other hand specific memories are not lost because they are not stored in one single region. Likewise, while language may be lost following a stroke, the ability to speak often returns because the brain's inherent redundancy means that other regions will always take over that particular task.

Another example of dangers in losing redundancy is provided by the tragedy of the Irish potato famine of 1845–1849. As already mentioned, traditional farming methods involve planting a number of crops and a variety of seeds for each crop. However, by the nineteenth century, much of Ireland's population had come to rely on the potato as its main source of food. What's more, only one or two high-yielding varieties were being grown. The result was that when the fungus *Phytophthora infestans* hit, most of the crop was wiped out. In just a few years, over a million people died as a result of this disaster.

Today, engineers and policy makers are learning important lessons about nature's built-in redundancy and robustness. During World War II Britain's seat of government was located in a bunker under Whitehall, London. At the height of the subsequent cold war, however, the United States realized that a distributed system of government was needed, one that would be resilient to specific damage. For this reason, straight highways were built that could be turned into emergency landing strips, and information was distributed through computer systems located all over the country. (It was in this way that the Internet was born.) Likewise, the control systems of aircraft and nuclear power stations have considerable built-in redundancy, through the provision of fail-safe and back-up systems.

Energy

Some experts predict that in 50 years the earth's resources of oil and natural gas will run out, or at least become so depleted that they will be too expensive to extract. It is as if we are now selling the family jewels at a cut rate so that we can run our cars, switch on our air conditioners, and heat our offices, homes, and shopping centers, while forgetting that our grandchildren may not enjoy such luxuries.

Half a century ago nuclear power—fission power—was hailed as the new energy source of the future. But a series of nuclear accidents have made people cautious about its safety. Of course, it is possible to design progressively more fail-safe systems in anticipation of various "worst case scenarios," but this does not always reassure the public.

Environmentalists, for their part, have proposed solar heating, wind power, and tidal power. These may help to augment other power sources but for many countries they will never provide a universal solution. Solar power may work well in Africa but it is less attractive in Canada and the northern United States. Tidal power can supply a considerable amount of energy in a number of places in the world but it is hardly practical in Switzerland! The combustion of biomass (for example, burning wood, or converting biomaterials into alcohol or methane) is another possibility. Yet, as the demand for food increases, there is going to be competition for land use.

The 1980s saw a big energy crunch. Oil prices rose and we realized that the world could be held at ransom by the oil-producing nations. The answer at the time was "conservation." But rather than making personal sacrifices in order to consume less energy, we preferred conservation without pain or personal inconvenience and, wherever possible, the help of a government grant. So we reduced our heating bills with insulation and weather stripping. Car manufacturers, for their part, came up with smaller models and more efficient engines that would continue to satisfy our desire for the automobile as a means of fantasy and escape.

In the end, this exercise in conservation was not totally painless. We learned that once a house is totally sealed against cold weather, air does not circulate so well. Radon gas coming up from the earth can be trapped in a house, along with the vapors produced by glues, plastics, paints, sealants, and, for example, Formica insulation. As a result, while we saved a little on our fuel bills we paid the price with allergies and diseases of the immune system.

When it comes to energy conservation there is no gain without a little pain. People have to realize that it is indeed possible to use less energy and that the sacrifice involved may not be too great. The houses in the Italian village where I live all have an electrical supply that, in terms of U.S. voltage, would be the equivalent of less than 40 watts. My previous house, in Canada, had a supply of 200 watts, which allowed me to indulge in a variety of energy-consuming devices that certainly do seem to make life that much easier. Now I am much more careful not to overload the house's limited supply. I don't leave electrical devices running, and I remember to switch off the lights when I leave the room. When I go to the small shops in the village I notice that the shopkeeper puts on a coat or heavy sweater instead of turning up the heat. And with hot weather in the summer, and a constant breeze in winter, no one has need of a clothes dryer. These may seem like tiny things, but multiply them by several million people and the energy saved would be considerable.

The future is uncertain. Maybe we will really have to tighten our belts and mobilize resources to meet the energy challenge. Some energy watchers believe that the remaining fossil fuels should be reserved for agriculture and essential industries. Some futurists argue that an energy crunch could be so serious as to require the mobilization of entire sections of the population, as is done during the crisis of a world war.

Who knows if, at some time in the future, we will be forced to give up our automobiles and join car clubs? Fast air transport will be an increasingly expensive luxury. One economist, Lothar Mayer, suggests that each baby should be given a smart card (an electronic chip) indicating how much of the earth's natural resources it has been allocated. It can then make a choice as to how to use that energy during its lifetime—on a car, a single trip in an aircraft, heating the house, and so on.

Global Warming

Around 10,000 years ago the peoples of the world faced a major climatic change as glaciers advanced to cover much of northern Europe, Russia, and North America. The result was a major population migration, followed by a return as the ice melted. This ice age also provided the opportunity for groups of Asian hunters to cross an ice bridge connecting that continent to Alaska. From there they spread into the American continent. (Other groups had probably colonized the Americas even earlier.)

There had been major ice ages before, during the Pleistocene era, just as there have been mini–ice ages in historical times. During one of the latter the Thames froze over to such an extent that Londoners were able to hold winter fairs on the river. Likewise, as the sun's output of energy changed, there have been periods of warming. We are now faced with yet another such period of global warming that could well result in major climatic disruptions. Again, this is the price we will have to pay for all that progress and consumption during the twentieth century.

A percentage of the sunlight that falls on the surface of the earth is reflected back into the depths of space. But naturally occurring gases in the upper atmosphere, including carbon dioxide, methane, and nitrous oxide, trap this reflected heat and direct it back to earth again. The effect is like the panes of glass in a greenhouse that cause it to be warmer inside than outside. Hence the term "the greenhouse effect."

Since the coming of industrialization and the widespread burning of fossil fuels—coal, oil, gasoline, and natural gas—the carbon dioxide content of the earth's atmosphere has increased dramatically. The natural consequence is global warming. Some scientists predict a temperature rise of 5°C over the next 50 years—taking us to the time when these same fuels will run out!

Five degrees does not sound like that much. In general, it will mean warmer winters and hotter summers. That doesn't seem a high price to pay, but the overall effect could be far more dramatic. To begin with, warming will not be uniform in each part of the globe; rather it will give rise to a series of localized but major climatic disturbances. When one area suffers from drought another will experience highly increased rainfall. Melting of the polar ice caps and mountain glaciers will release an enormous amount of fresh water and result in a rise in the levels of the world's oceans, causing flooding of coastal areas and the possible inundation of coastal cities. Also a combination of climatic change and the influx of less dense, fresh water may cause the flow of ocean currents to be modified. One major concern is the Gulf Stream. If its direction changes, then while the rest of the world heats up, Northern Europe could be plunged into a mini–ice age.

A further concern is the rapidity of this warming. Balanced ecologies of animals, insects, plants, and trees are tolerant to small climatic changes. But if the temperature rises too quickly and too high, entire ecosystems will be threatened. Many plants seed themselves each year and so it won't be too difficult for them to migrate and set up niches in cooler regions. The same, however, is not true for trees. Climatic change in the past has been sufficiently slow to allow time for trees to move to more appropriate regions. But a temperature rise occurring in just a few decades could wipe out some major forests and ecological resources.

The world's governments have been alerted to the possibility of global warming and are talking about ways to reduce the amounts of carbon dioxide released into the atmosphere. Only time will tell how effective these measures, if carried out, will be.

Conclusions

This chapter has only touched on just a few of the issues, dangers, and "disturbances" that face each one of us and our collective life on this planet. We could have explored the issues of genetically modified foods, the crisis over water supplies, how to dispose of nuclear waste, and a host of other issues. But in case after case the overall principles involved are more or less the same. We face a variety of ecological and environmental issues often brought about by human carelessness and thoughtlessness. We are constantly producing new technologies, new materials, and new products, yet we can never predict with 100% certainty what their impact will be on society and the environment. Some information will always be missing for us. Some risks will be unforeseen. Some implications may be more significant than we expected.

What can we do if we can't have complete control over the world around us? What is the point of plans and policies if we can't really know the future? Had we been blessed with hindsight would we have done anything to avert the crises we now face?

The answer, I believe, is to look to nature's own model. Life survives on this planet in a wide variety of highly improbable locations—deep in the oceans where no light penetrates, within volcanoes, inside nuclear reactors. Life has survived major climatic changes in the past. Diseases, which are one expression of life's versatility, find ways to subvert our antibiotics and disinfectants.

Nature always wins because of its profligacy and abundance. Nature survives for the very reasons that would entirely frustrate traditional businesspeople—it makes endless duplications and is replete with redundancy. On the face of it nature appears hopelessly inefficient and disorganized. It takes only one sperm to fertilize a human ovum yet a man produces hundreds of thousands at each ejaculation.

Look at any roadside and observe the vast number of different grasses, weeds, and flowers. Pick up a handful of earth and note all the tiny organisms that are scurrying around. Nature is abundant. Nature overproduces. Nature is not content with producing one type of bird, fish, or plant but explodes into endless varieties. Lift a rock, walk in the woods, poke at a dead tree and you will find ecosystems within ecosystems all joined together into complex networks of symbiosis and mutual support.

This is why nature survives. Nature does not produce just one variety of apples, potato, or wheat but a multiplicity, and while one variety may be vulnerable to attack by a particular disease, others may be resistant. When fungi, disease, or insects attack a farmer's field or orchard, a year's crop may be wiped out. But if that same field is allowed to grow wild, or with a broader range of fruits and crops, then only a percentage of its growth will be destroyed.

Nature keeps its options open. Nature covers all its bases. Nature makes not one master plan but many. Nature does not have an exclusive policy for the future. And this is where we can learn a great lesson. Of course, we must still make plans and policies, but as we make them we must acknowledge that certain elements in any situation lie beyond our control and that no plan or policy can be comprehensive or take into account all possibilities for the future. Our policies and our organizations should have the same built-in flexibility that is exhibited by nature.

Ironically, in the last analysis we all need to be that much more inefficient. Maybe it is a good idea to have one or two dreamers in every business—they may not produce as much as the others but one day their intuition and vision could be important. Maybe it is useful to have the odd eccentric around, someone who doesn't quite fit in because his or her thinking is different from everyone else's. Yet, when a situation changes in an unexpected way it could be just that same offbeat thinking that saves the day. Maybe not everything should be accounted for in an organization. If it is going to be capable of adjusting to change then it must have room to maneuver.

Human ingenuity and human creativity are limitless. We need our organizations, our governments, and our approaches to the future to reflect just a little of our inherent genius.

Eight

PAUSING THE COSMOS

A Dream of Enlightenment

Americans felt confidence in their world at the birth of the twentieth century. The decades ahead would be unperturbed by the uncertainties of international politics, for America still adhered to the Monroe Doctrine of 1823, which declared the Western Hemisphere closed to further colonization and expressed the U.S. policy of nonintervention abroad. An international peace conference had been held in The Hague in 1899 and a year later the United States adopted the gold standard so that its paper money would always be backed up with something tangible.

In that same period British children were taught that all those red areas on the map of the world were British colonies and protectorates. The British Empire, they were told, was much vaster than any empire in the history of the world. It literally spanned the globe, so that the sun would never set on its boundaries. How could such an empire, based on trade and paternalistic administration, ever falter?

Americans and Europeans alike were inheritors of the great Enlightenment dream whereby people could be improved and society bettered through knowledge and education. The eighteenth century Enlightenment philosophers had expressed their confidence in the power of reason and the value of progress. They believed it would be possible to eliminate extremes of poverty and inequality. Cities would be orderly, rational places. And, once they had been freed from want, human beings could be counted on to act in the best interests of those around them and treat others as they would wish to be treated themselves. If crime and antisocial behavior were the result of poor housing and faulty social conditions then such ills would be eradicated by rational social planning. With a well-educated and properly informed public, true democracy would be possible.

This dream was based on a set of collectively held certainties, values that everyone espoused—the common good, maximum happiness, reason, free will, good government, and the rule of order. It had its seeds in the city-states of the past, from Athens of classical Greece, to Florence and Venice of the Renaissance.[1]

City-states were small enough, and sufficiently compact, for a vibrant democracy to be practiced (although suffrage was by no means universal). A small group of elected officials, responsible to the whole society, could act in an enlightened and responsible way and make wise and sensible decisions to give society its internal stability and protection from outside disturbance. The citizens of such states were both content and creative. Not only did they practice trade, but they also had a love for art, music, literature, and beautiful public buildings. The artist Piero della Francesca, for example, drew up plans for an Ideal City, for, after all, rational people should live in rational spaces. In turn, a city founded on mathematical principles would induce harmonious and orderly behavior in its citizens.

---

[1]This is not to say that other peoples, from the Shang of Ancient China to the Blackfoot and Iroquois confederations in North America, did not also organize themselves wisely.

Even the Dionysian elements of human nature were not ignored by such rational societies. Room was made for them so that they did not erupt in an uncontrolled way to threaten peace and order. Venice and other city-states had their periodic carnivals, during which sexual license was permitted, but always within a framework that would contain rule-breaking. By hiding their faces behind masks, for example, anonymity was preserved so that family relationships would remain uncompromised. When, in the sixteenth century, Venice experienced a rise in the number of male prostitutes, the city avoided any confrontation with the rules of the Church regarding homosexuality by decreeing that a man wearing a female mask was officially a "masked woman" and therefore free from arrest. The forces of human desire were thereby contained through the exercise of wisdom.

The Enlightenment had turned its back on superstition by stressing that "man" is a rational animal. Then came the Utilitarians, John Stuart Mill and Jeremy Bentham, who argued that it was possible to maximize human happiness, just as it is possible to quantify and maximize any other commodity. The eighteenth century also saw the rise of science and an increased faith in the power of knowledge. Its logical outcome was the belief that science and its associated technologies could solve the outstanding problems faced by society. Such problems would be approached in a "scientific way" using logic, knowledge, and the ability to predict the future through mathematics.

The faith in a scientific future reached great heights with such technological optimists as H. G. Wells (although Wells could also see science and the future of human society in a pessimistic light). Human beings would discover inexhaustible sources of free energy; they would live longer; disease and famine would be eliminated; all knowledge would be revealed to us. Thanks to rapid communication and ease of travel we would realize that we inhabit a single world, and wars and conflicts would be things of the past. Poverty and inequality would be eliminated and there would be a world government of benevolent technocrats. This was the image of the future at the dawn of the twentieth century.

The Sleep of Reason

Yet, in the years that followed, two world wars erupted along with countless other armed conflicts, mass repressions, ethnic cleansing, the Holocaust, germ and chemical warfare, environmental devastation, and the threat of nuclear annihilation. Such events left many thinkers in a state of shock. Artists, composers, and writers asked how it would ever be possible to make new works in the shadow of such horrors. How could they express beauty, joy, confidence, and hope in the light of everything that had happened? Even science had become tainted. In the words of Oppenheimer, with the exploding of the atomic bomb science had known "original sin." Supposedly decent people—physicists, chemists, engineers, and psychologists—have devoted their talents to the production of weapons of mass destruction—nuclear bombs, rockets, poison gases, germs, and viruses, as well as the means to brainwash, torture, and destroy the human personality. Politicians, bureaucrats, and generals have drawn up plans for mass annihilation, the destruction of entire populations, and ethnic cleansing.

The greatest horror is that, after the devastation of two world wars and the constant threat of nuclear annihilation of life on earth, the old patterns of thought continue. Disputes are still resolved by violence and war. In some cases violence is brutal and direct, as with the rape and butchery of populations, in others it takes advantage of high technology to deliver death at a distance with rockets and electronics. It seems that even our most sophisticated science and technology are being put to the service of our most primitive drives and reactions.

Despite all our new knowledge, our science, our international courts, the United Nations, and our ability to communicate globally so that "nation shall speak peace to nation," the terrible mess continues. Does this mean that reason and science are insufficient? Is the human race an evolutionary experiment that is now failing to the point where it could well destroy both itself and the environment that supports it? Did human consciousness develop too rapidly to deal with the technologies it created? As moral beings, are we doomed to lapse again and again? Is there any way we can be saved?

When destructive behavior is observed in others we turn to the psychologist and psychiatrist for a diagnosis. Freud argued that reason, the supposed firm foundation on which a rational society is based, is no more than the surface of a vast ocean of the unconscious, a hidden region of impulses and desires. The forces of this unconscious constantly threaten to break through into our waking life.

Humans are driven by the forces of Eros—the libido or life instinct with its drive for pleasure, sexual release, and survival. But there is also a counterforce, that of Thanatos, or the Death Wish, that seeks resolution to all of life's tensions by returning to an undifferentiated, inanimate state of death. Thanatos and Eros form a duality in constant conflict within individuals and societies.

Toward the end of his life in 1930, while suffering from cancer of the jaw, Freud explored the conflict between Eros and Thanatos in his deeply pessimistic *Civilization and Its Discontents.* Freud ascribed the oceanic feeling of oneness, common to mystic states and at the heart of all religions, as a desire to return to the helpless infantile state of total identification with the mother. He believed that there could be no resolution of this yearning for a return to an undifferentiated state. Unlike the Oedipus complex, which can be worked through in psychoanalysis, there is no "cure" for this desire to resolve all tensions through death. It is common to all human beings and underlies all societies, where it constantly threatens to erupt into profoundly disruptive behavior.

The tension between civilization and nature (in the form of our deepest drives) is irresolvable. According to Freud, it is for this reason that there can never be a truly ideal society or unalloyed human happiness and harmony. It may be the case that extreme poverty, social injustice, and political inequality trigger outbreaks of violence and tensions. Yet the deeper origin of such tensions lies with Thanatos, the death wish that is projected outward onto nations, racial groups, and individuals. Because Eros and Thanatos are irreconcilable, human guilt and the absence of total happiness are inevitable. All forms of civilizations are, at their core base, a thwarting of our most basic drives and desires.

Much of Freud has been discredited, or subject to a radical rereading, yet his argument, that there is a deep conflict between the desire for civilization and the underlying drives of human nature, is highly persuasive. What other meaning, than a projection of Thanatos, could the symbol of the mushroom cloud that hung over the world for decades have? What were those generals doing with their war games as they talked about megadeaths? Why did scientists devise a neutron bomb that would destroy human beings while leaving their buildings intact? And, now that the nuclear threat has to some extent been defused, why do we look up into the skies for an asteroid or giant meteor to be the new bringer of death? Armageddon, we are told, will arrive from the stars and smash into the earth, creating great tidal waves and dust clouds high in the atmosphere that will block sunlight for years and produce the equivalent of a nuclear winter.

The very opposite of this desire for death should surely be the drive toward life and a passion for the natural world. Yet even within the environmental movement itself one can find hints of Thanatos. It exists as the fantasy of a major ecological disaster that will wipe out human civilization (a variant of the Native American story of the Great Cleansing).[2] Much of the human race will be destroyed as nature fights back, and only small, simple communities of like-minded people, caring for the earth, will be left. In this sense, while the environmental movement is motivated by the highest ideals, by its love of the natural world and the right to life of all species, it is also associated with a profound sense of guilt at being human—another factor in Freud's analysis of the human situation. In some instances this can erupt as anger and rage against those who are perceived to be the various enemies of the natural world. Guilt and rage, combined with a desire to resolve all tensions through a metaphoric form of death, can contaminate ecological thinking.

Thanatos has been manifest in the mass suicides among religious cults. It is present in extreme political groups that stockpile weapons in

[2]When I write of the "fantasy" of ecological disaster or an asteroid impact I do not mean to suggest that such an event could not occur. I am using the term more in the psychological sense of an event, real or imagined, that becomes a focus for emotionally charged acts of imagination.

anticipation of Armageddon and in the rising number of penal executions, with their tendency to become media events.[3]

For Freud the only possible strategy is one of resignation and acceptance of the inevitability of human nature. Civilization will never achieve all it pretends to. Reason has its place—certainly it cannot be abandoned in favor of some instinctual and intuitive reactions to things. Reason helps to put a break on our more violent impulses and poorly thought-out reactions but, according to Freud, it can never be enough. It only goes part way and is never strong enough to overcome our more basic drives.

Freud offered a psychological interpretation of the human situation based upon the duality between Thanatos and Eros. A few decades ago an attempt was made to provide a biological explanation. It pointed out that the human neocortex, that advanced part of the brain capable of language, reflection, and planning, is a relatively recent evolutionary product. Anatomically it is grafted onto an earlier mammalian brain and the more ancient reptilian brain. Thus the human brain is triune, with three structures, one superimposed on top of the other.

The recently developed and somewhat immature neocortex is responsible for controlling the less "rational" drives and impulses of these earlier brains. Like an inexperienced schoolteacher with a class of unruly children, it is not always able to control their underlying outbursts. On the surface this again looks to be a good explanation for the limit of human reason to control our more "animal" natures.

Freud and the triune brain theory point in similar directions. Human evolution is far too recent for societies to exist in stable form. The neocortex and superego are simply too weak to control the brain's underlying forces and drives. Reason and civilization are not sufficient to overcome the conflict between Eros and Thanatos. Human beings, at their present stage of evolution, are flawed. In view of the rapidity with which science and technology can produce the means of mass destruction, the future of the human race, as well as that of the other organisms who share this planet as their home, looks pretty grim.

---

[3]It even seems that some executions are timed to coincide with prime time viewing.

But is this analysis complete and must its pessimistic conclusions be accepted as the ultimate explanation for the human condition? It seems to me that yet again it is an aspect of the Enlightenment dream, but this time turned into a nightmare. It suggests that human beings are progressing somewhere, that they have come so far, but are not yet sufficiently strong in their control of unreason and animal instincts. It is a view that seems to go back to the early Church Fathers and their desire to subjugate the flesh. It derives from the notion that the world is somehow evil so we must purify the spirit to the point where it can leave the body behind. It is here, I believe, that one encounters most forcefully the Western mind's deep sense of guilt—at pleasure, at the body, and at our desires.

But pause for a moment to look at the animal world. A dispassionate observer will not see "nature raw in tooth and claw" but a balance of nature and a circle of life. It is true that some animals graze, gathering together for mutual protection, while others hunt them for food. Yet hunting animals do not kill indiscriminately. Wolves pick out the weak and sick animals in a herd and kill for their own immediate needs. In this way a balance of life is maintained by weeding out the sick and weak and avoiding the overpopulation of any one species.

Neither do the members of a species turn on each other and kill—unless they are kept in artificial or highly confined spaces. Dogs growl and leap at each other but mainly this is a form of mock fighting, a highly controlled form of display in which blood is rarely drawn. A pack of wolves show less aggression amongst themselves, and far more self-control than a group of head-butting teenagers emerging from an English pub on a Friday night. Even the most basic drive, sex, is sublimated within the animal kingdom into elaborate rituals of courtship.

The briefest glance at the animal world should tell us that "animal instincts" are stylized, geared to the good order of the pack, and to the sustenance of the entire balance of life. Judging from what anthropologists have found in various areas of the world, the earliest hunter–gather groups also lived within the balance of nature. It is true that when two groups were forced to share, or to hunt, within the same territory, acts of aggression and even warfare occasionally occurred. Yet in several cases such societies learned to sublimate their aggression into ritualistic forms such as the exchange of elaborate gifts or a symbolic game. Likewise, within a small society in which the members sit around a fire or in a meeting hut and discuss and address the various tensions as they arise, violence and disruptive behavior can be contained.[4]

Without idealizing early and indigenous cultures, it appears that in the main they were relatively peaceful and offered no major threat to each other or to the surrounding environment. The conclusion I draw is that, left to themselves, our animal drives and instincts are not that harmful or destructive, and human reason is quite able to deal with them. Reasonable people can postpone immediate gratification in order to reach more important goals. They are driven more by the desire to help and cooperate than to compete in destructive ways.

It is not so much our underlying drives and instincts that are creating problems in the world as our higher functions—reason, imagination, memory, and so on. Higher functions enable us to build up pictures in the mind, to engage in fantasies, and to revisit memories and clothe them anew. Our higher functions enable us to abstract aspects of the world and treat them almost as objects or models. Just as we can build a toy car or train and turn it around in our hand to examine each aspect, so too we can create an idea, a concept, or an image in our minds and manipulate it like an object. We can also externalize the objects of our thought, projecting them outward onto others.

---

[4]The Cassowary Cult of the Pacific Islands is one example whereby rivalry and competition became ritualized into the annual giving of Cassowary birds between groups. The bird is valuable and so prized that a group must focus all its energies throughout the year to obtain these birds, which are then given away as gifts. The exchange of gifts also establishes a mutual web of obligations, which can cement potentially rival peoples together.

The potlatch of the Pacific Northwest is a further example, in which the head of a family hosts a great feast and gives away extremely lavish gifts. In part this establishes status, in part it can ritualize potentially dangerous rivalry.

The Palio (horse race) of Siena originated in medieval times as a competition between the districts or *contrade* of the city. While it has become something of a tourist spectacle, for the Sienese it is very much a living ritual with enormous rivalry occurring in the days leading up to the race and days of feasting afterwards.

Our thoughts are like a great stage. We people this stage with characters and endow them with emotions and goals. We forget that they are no more than the products of our thoughts, that they are smoke and mirrors. Nevertheless we end up treating them as if they were real, autonomous things in the world. This is where our problems arise, not in our "animal instincts" but in the distortions of reason whereby we become incapable of distinguishing the products of our thought from those of real objects or situations in the world—and of course real objects are also, in a sense, created out of our perceptions. We do not so much see the object in all its naked reality as we see, in part, what we expect to see.

We spend parts of our lives out of contact with what could perhaps be called "the real." We don't always live in the present moment. We are disconnected from events. At the self-same moment that we are experiencing something, we may also be standing outside ourselves observing our reactions. At a moment of pleasure we may already be in the future anticipating the next occasion. Being in one place we may imagine ourselves in another.

The brain is exceptionally creative. It is able to summon up dreams and images to the point where they end up creating a half fantasy world where nothing is really immediate. The fictions of our thought become realities—enemies, foreigners, evil powers, economic threats that literally threaten to destroy us. And that word "literally" is chosen because what is under threat is the entire theater of our thought, a construct that has become so real for us. If it were to collapse, then we believe we too could disappear along with it. In the face of such threats we must either fight or flee. And so we no longer relate directly to people, events, and situations around us but focus on the Other that we have created in thought. This Other may be a person, or a particular group of people. It may also be some perceived threat to our existence or well-being—violence in the inner cities, environmental damage, and the spread of drugs. It is not that such threats and dangers do not exist in actuality, but that they have been amplified and clothed by thought to the point where they become monsters of the imagination so that we can no longer distinguish the products of our own thought from what lies outside in the world.

It is our higher functions that are hijacking our deeper instincts, rather than vice versa. We are not so much the innocent young teacher incapable of controlling an unruly class of instincts. Rather we are the paranoid teacher who is inciting the class to violence, by portraying some vast threat, some fear, some Other that must be opposed at all costs. It is the perversion of thought and reason that poses the threat to our civilization, not deeply buried instincts or asteroids from outer space. The mushroom cloud of the atom was the creation of human reason. The "evil face of communism" that threatened at any moment to end all life on earth was the nightmare of reason, not reason's logical analysis. On both sides of the Iron Curtain we lived in fear of the images created by our own thoughts. We were like little children frightened by our own shadows cast on the bedroom wall. These monsters of reason had no more substance than ghosts. Yet in their name human societies are willing to rape and murder, to bomb and release wholesale nuclear destruction.

It makes sense that Freud's Thanatos, that deep wish for death, is projecting itself into the heavens in the form of an earth-destroying asteroid. But what complements this death-wish is the amplifying and distorting power of human thought and reason. The human imagination clothes this image of death from the heavens. It gives it substance and expresses it scientifically. In this way a death-giving asteroid becomes very real in the human mind. It expresses itself in the fantasies of cinema, such as the movie *Armageddon*. It gives the impulse to those groups of scientists and amateurs who are now monitoring the heavens for approaching asteroids, and to politicians who plan even more powerful nuclear weapons to attack this threat from space. The underlying drive may be primitive but it has been elaborated and turned into a scenario of human thought. That is where the true danger lies.

Thought and the creative powers of the human mind have produced our modern world with all its triumphs, technology, and discoveries. They have also produced wars, violence, and environmental devastation. Human thought has always been this way, since prehistoric times. But in the distant past groups were small enough to contain inner tensions and, even more importantly, technology was not so advanced. Now that the power of our technology increases without limit we need wisdom if we are to put our house in order.

Taking Stock

Where are we to find this wisdom and how are we to use it in the world? Think of artists working on a portrait or still life. From time to time they stand back and squint at the canvas. They have been working on a particular detail and now need to pause and look at things in perspective, taking the whole vision into account.

Something similar happens in all creative work, as well as in our daily lives. We can get so tied up in details, in rushing toward a deadline, in the routine of the office, in the need to make profits or attain promotions, that we forget to pause and ask, Just what am I doing? Why am I doing this? Is this really what I want? Is this really what I set out to achieve five or ten years ago? Is my life fulfilled? Am I happy and content in my family relationships? Do I still hold to the ideals of my youth? Or have I been caught up in some grand scheme, sold a bill of goods, made compromises, bought into a system in which I no longer believe?

We should be asking such questions throughout our lives. But more often they are only associated with what is termed the "midlife crisis" when people start to question their lives and at a time when they have a sufficient perspective to look back upon the path they have taken.

What is true for individuals can also be true for a human society. Maybe it is now time for Western society to take such a pause, to step back and ask: Where are we heading and where have we come from? What impact are we having on the world? Is society some abstract ideal or is it us, we human beings who are its members? And if so then what are we doing to ourselves? What are our individual values and how do they resonate with the values of the society in which we live?

It is indeed time to pause and step back from the painting. It is time to ask ourselves if society needs to rush ahead at such an accelerating pace. The events of the twentieth century caused us to question the Enlightenment dream and its assumption about continued human progress. Yet if we leave this dream behind then where are we to find the meaning upon which our society is to be based? This is the key issue we face at the start of our new millennium.

Every society has a foundation in shared values and meanings. Many indigenous peoples picture themselves as caretakers of the earth, with the natural world as a gift left in trust to them. Some societies are concerned with maintaining a living connection with their stories of origin. For them continuity is more important than change. Groups have also been founded upon the ideals of compassion and love. What will be the deep meaning of our new world?

To a greater or lesser extent the values of Western society, the values that have brought us to the dawn of the third millennium, have been based upon the ideal of progress; not only progress in terms of the accumulation of wealth, property, goods, and knowledge but also social and human progress and its continued evolution. In these latter aspects we again see the Enlightenment dream that human beings can be "improved" in a variety of ways.

It is perfectly natural that change should take place in human affairs as it does in the natural world. The problem arises when this change is understood only in a one-dimensional way, as related to progress and, in turn, progress as something that can always be quantified and commodified. In this way progress becomes a goal in itself. It is something we always need more of. Change in itself, rather than its particular content, is measured by the ways it contributes to, or delays, progress. But to view human culture with its art and music, its religion and human psychology, only through the perspective of "progress" is to impoverish our experience of the world.

Notions of progress so permeate our Western way of thinking that it is difficult to view the history of any subject without adding the gloss of a linear ascension in time. Now the moment has arrived to suspend our immediate desire for progress and examine the whole structure of the society we have created, and the direction in which our world is moving.

The Vision Changes

One step in that direction is acknowledging that our world is more complex than we ever imagined. In that sense, ultimate explanations and totally objective observations may not really exist. Science has begun to set aside the blinders it has been wearing for the past 200 years to view the world in terms of complexity, ambiguity, and uncertainty. If the material world appeared simpler in the past it was because we were looking at it through the perspective of classical physics. When we choose to direct our sight only toward simple systems (for example, those close to equilibrium or that are acted on by small forces, and that behave in regular ways) then naturally the world appears simple. It is a little like those travel brochures produced several decades ago by the apartheid government of South Africa. A naive reader could be seduced into believing that the population was overwhelmingly white because only white faces were seen in the carefully posed photographs of shops, bars, and beaches.

Similarly, classical physics created a travel brochure of the cosmos, one that emphasized regularity and simplicity. Galileo idealized his observations of the way a ball rolls downhill by ignoring, or bracketing out, the effects of bumps and friction. Newton asked how an apple falls in the absence of air resistance. Chemists investigated reactions where everything was close to equilibrium. Scientists were interested in what they termed "closed systems," systems insulated from the perturbations of the outside world. When it came to the study of solids, such as metals and crystals, they developed theories about tiny disturbances, small vibrations, and gentle heat flows. In each case science was filtering the world. And because theories of closed systems, reactions close to equilibrium and small disturbances, worked so well, scientists naturally concentrated on investigations within the context of those particular conditions. Carefully designed experiments, well insulated from the contingencies of the external world, provided clear data that would fit easily onto a graph without too much scatter or experimental error.

The world of classical physics was free from uncertainty, ambiguity, and chaos. In turn, scientists set up their experiments in ways that confirmed these basic assumptions about the world. This is what Thomas Kuhn calls a scientific paradigm. Science always works within paradigms, which means that new knowledge is always gathered from within a particular context and by making assumptions that are held largely unconsciously. The result is that such knowledge naturally falls within the established scheme of things. It is only when science comes to a barrier and can go no further that a paradigm begins to break down. That is when a true scientific revolution becomes possible.

During the early years of the twentieth century physicists struggled to integrate the new discoveries about atomic spectra, quanta of energy, and the structure of the atom. In case after case, their thinking was confined within the paradigm of classical physics, while at the same time making some modification to existing theory. Even Niels Bohr, in his first attempt at an atomic theory, grafted new insights about quanta onto the old idea of classical orbits. It was only when Heisenberg broke with the traditional way of seeing things that modern quantum theory was born.

The same applied to the anomaly of the orbit of Mercury that violates Newton's laws of motion. In their desire to hang on to the Newtonian paradigm, physicists attempted to account for Mercury's orbit in terms of gravitational perturbations arising from irregularities in the shape of the sun. It was only with Einstein's revolutionary idea of relativity that this problem could be resolved and incorporated into a new way of thinking.

It is always possible to save an existing theory by grafting on more and more assumptions and corrections. At the time of Copernicus, for example, astronomers were still trying to save the Ptolemaic earth-centered solar system by adding in epicycles within epicycles. In the end these corrections became so messy and arbitrary that it was clear that a revolutionary new gaze was needed.

As we move into this new century we realize we have been guilty of oversimplifying the world in so many fields of knowledge. We have been looking at nature and ourselves through the convenient lenses of theories that present the cosmos to us in limited ways. Now we acknowledge the inherent restrictions of any theory. We recognize that nature is complex in its details, unpredictable, and often uncontrollable. What is true for the natural world applies equally to human beings and their societies. It is for this reason that our entire society needs to pause. Notions of continued human progress and development must be carefully reexamined if society is to be founded on wise values and enriching approaches.

Reading Character

Our realization of the inner complexity of the cosmos and the multiplicity of approaches needed to understand the world is not confined to the world of matter but also applies to our own inner universe and ourselves. At first sight it would seem that, as certainties dissolve into ambiguity and uncertainty, the one thing we can hold onto is ourselves, our sense of identity as independent beings in the world. We may apply philosophical doubt to the world around us, but at least we are assured of our own existence. Descartes doubted everything but could never dismiss the thought that was constantly doing the questioning. Everything can be subject to doubt, but what about the doubter himself? The fact that he questions must in some way confirm an existence. The observation that thinking is going on must imply a thinker. Thus Descartes came to his famous conclusion "I think therefore I am." Let us cast this in another way: "Thinking and questioning are going on, therefore there must be a thinker and a questioner." To jump from this latter statement to the deduction that there must be an "I" as an independently existing being with a past and a future, an existence in space and time and a well-defined personality, is more difficult to justify. One can say that an activity of thought is going on and that questions arise in search of answers. But does that necessarily imply the persistent existence of a thinker who has a clear continuity from past to future? The action of thought is quite different from the existence of a rock, for example.

Each of us has a birth date, telephone number, Social Security number, and other means of identification, but are we really clear as to who this "I" within us really is? We have memories, some of which change as we age. As we look back we see a child or teenager, sometimes with different behavior, beliefs, and values from what we would now hold or practice. Our name is a point of continuity for others but how much else remains the same over the years? And is the "I" it names really more fundamental than the various personae we present to the world?

The term "persona" derives from the Greek word for mask. In many theatrical traditions to don a mask is to become a particular character or stereotype. In the early Italian dramatic form known as commedia dell'arte there are the stereotypes of the doctor, soldier, and so on. A mask presents a character with graphically etched characteristics who behaves in ways consistent with those traits—miserly, overbearing, officious, seductive, or foolish.

As we enter an office or workplace we also put on a mask or persona. A person "becomes" a teacher, police officer, doctor, store clerk, airline pilot, waiter, and so on. The uniform and the setting help to set the scene. And "scene" is a good word to use because entering a workplace is a little like stepping onto a stage on which various scripts are about to be played out. Visit a doctor's office, or go into a bar late at night, and you may end up talking about life's intimate details. This would not be the case in an encounter with the maitre d' of an expensive restaurant. A man expects his barber to discuss details of last night's game but this sort of conversation would seem singularly inappropriate when asking a bank manager for a loan.

Specific situations and uniforms call for specific scripts, "speech acts," and personae. Step out of that situation and you become a different person. Leave the bank and you cease to be manager. Join your friends at the bar and you become "one of the gals." Put your key in your front door and you become Mommy, Daddy, husband, wife, or partner.

Putting on the persona of maitre d', nurse, bank manager, doctor, or schoolteacher may, in part, be a relatively conscious action. A person puts on the uniform, steps into the office, consulting room, bank, or restaurant and, to a certain extent, acts out a role. One is saying in effect, "this is what I'm good at, this is my profession, but this isn't really me. Behind this mask I'm quite a different person."

Behind that mask lies the wife, husband, lover, parent, or child. But are these also personae? Am I a father or am I "being a father" by putting on a particular act for my children?

In our lives we wear a variety of masks, some consciously and some unconsciously, to the point where some of us are no longer sure which is the mask and which is ourself. Take as a particular example an actor friend of mine who found himself cast in the role of a contemporary Don Juan. Over the weeks that followed he began to change. He was more confident and outgoing and eventually had several brief affairs with women. After the play closed his behavior gradually slipped back to his earlier self.

An actor puts on makeup and particular clothing, adopts body postures and gestures, and speaks a particular script. In this the character created by the actor is a little like a persona. Likewise, some people "act the part" of a maitre d' or bank manager in their daily lives. Many good teachers say that what they do is close to a "performance" in front of their class.

In most cases these masks can be taken off and put on with each performance or situation, but in the case of our actor something slid over into his daily life and stayed with him for a time. All actors don't have that problem, yet with particular powerful and dark characters, such as Lady Macbeth, there is the danger that the character will "take over" some aspect of a person's individual personality. Or conversely, in order to play such a character, an actor must discover aspects within him- or herself that have lain unsuspected for years or even decades.[5] The point I'm making here is that the persona can eventually become ourselves, or an aspect of ourselves, to the point where we don't know which is the mask and which is the I. And so we ask: Is there really a central, real, and true person? Or are we all a complex series of aspects and creations? Rather than the "I" being a stable object in space and time, is it more like a process or integrating principle that collects ever changing fragments together and binds them, for a time, into patterns of behavior, attitudes, and motives?

This is analogous to the processes of vision, which begin with various strategies for seeing—edges, bars, moving fields, patches of color—that operate relatively independently and are only later integrated into a tree or a face. So too, the self may not be so much a fixed object as a

[5]Almost paradoxically, the worst villains must be played with sympathy otherwise they become cardboard figures. Sir Alec Guinness spoke of this dilemma when playing Hitler in the movie *Hitler: The Last Ten Days.* He had to delve both into himself and into the character to discover something of sympathy that would engage the audience and provide them with a motive to re-create this character in their own reading of the film.

series of hypotheses and uncertainties, costumes, masks, and personae with which we face the world. The more the world reflects these back to us in a confirming way, the more we act consistently.

Deep within the rigid and formal teacher there may be a child crying for release. Within the seductive vamp, who is always breaking hearts, there may be a warm and loving mother. Who we are and how we appear to the world is always filled with paradox. Being ourselves is like Cézanne painting a landscape—he who was always tentative, always questioning, never fully sure but always attempting to respond honestly to his "little sensations" as he called them.

Another clue to the extent to which we have the ability to create a persona comes from the way we "read" and thereby create a character in a book. A traditional novel invites us to suspend our disbelief that we are reading a work of fiction and to imagine that we are following what is actually happening to real persons. Characters have lives, and their past histories begin before the novel starts. Their various encounters are located in a real place and time. After the novel has ended these characters go on living and their relationships continue to unfold. Victorian novels generally contain a concluding chapter that ties up loose ends, explaining how a particular character eventually married and had children. Villains get their just deserts and, toward the end of their lives, repent and make amends for the harm they have done. Some characters even continue to live on to appear in other books. In Jean Rhys's *Wild Sargasso Sea,* Mr. Rochester's wife, from Charlotte Brontë's *Jayne Eyre,* has an existence in the Caribbean before she marries and moves to England. The school bully, Flashman, of *Tom Brown's Schooldays,* grows up to engage in a series of picaresque adventures written by George MacDonald Fraser.

In one sense this is perfectly reasonable. Characters do come alive for their authors. They take on independence to the point where their author is constrained as to how far events involving that character can be pushed. Authors can even be surprised by what their characters do or say. Some characters insist on returning in subsequent stories or novels. Anthony Burgess, who wrote several novels about a costive poet named Enderby, once experienced the temporary hallucination of seeing his creation sitting on a lavatory and writing!

Where, then, does a character exist? In the mind of the author as some private image that must then be set down in writing? On the page as a unique objective reality? Or in the act of reading itself, in which each reader is creating something new and different?

Think of a male reader who picks up a work of escapist fiction, one of the successors to James Bond for example, a hero who fearlessly battles foes, drives fast cars, and can dismantle nuclear weapons or hack into elaborate computer systems. The male reader loses himself in the book, fantasizes about a world in which he too has indomitable courage, endless endurance, fast reflexes, and instant success with women. For a time the reader becomes that character, then the book is set down and the real world returns.

Suppose the same reader now picks up *David Copperfield.* Again a form of identification takes place, but on a more subtle level. The early parts of the book deal with the pains of childhood, the love of a caring mother, the brutality of a stepfather, and a sense of being cast out into an adult world. The reader sympathizes with David and recalls both warm and painful instances from his own childhood, perhaps his first days at school, bullying, a friend he admired, or an early love affair.

The character David is clothed in a more subtle way than the stereotypes of a spy novel. Each reader creates a different David Copperfield. After all, the reader may be English or American, an only child or one of a warm family of brothers and sisters. Copperfield is no longer a character confined to a book but has aspects of a real person, a person who has been brought alive through the creative act of reading.

Suppose the reader is a woman. For a time she suspends her femininity to identify with a male child and through that child's eyes she sees a mother and the nurse Peggoty. A woman, suspending her disbelief, enters the world of a young boy and in turn brings to life female characters that have been created by a male writer.

Even more complex would be a Victorian woman's reading of *Wuthering Heights*, a novel whose author was originally listed as a man, Ellis Bell. In it she meets the passionate Catherine Earnshaw, a fully rounded female character far from the shy and delicate heroines of a Dickens novel. Yet Catherine's story itself is told through two observers, a male visitor to the house and a woman servant who unfolds the history. The female reader has to perform a highly complex act of creation and interpretation as she brings to life a woman wild with passion, driven to extremes, and identified with the wild Yorkshire moors. What a shock when our Victorian reader later learns that Ellis Bell is the alias of a woman, Emily Brontë. Suddenly the entire novel shifts and dislocates and Catherine Earnshaw must be read and re-created anew.

The act of bringing a character to life is of necessity performed though the context of our own cultural assumptions. The way we read is always within a context of age, ethnic origins, family history, sex, sexuality, and education as well as all the books we have already read. Each time we pick up a book it is different because we have changed and we are bringing something new to the act of creation. When a film or television series is made of the book we may say, "That's not really Heathcliffe," or "That's not the way I see David Copperfield." Each actor will create a different Lady Macbeth or a different Hamlet. Each director will reanimate a play by Shakespeare and find within it something entirely topical and apposite for his or her own time.

What applies to characters in a book, I am arguing, is equally true of the ways we are constantly creating ourselves, integrating our various personae and attempting to connect to what we feel is our essence. But is there really a "true essence"? Is there an objective aspect that remains constant through time? Or are we more like open systems and processes within the constant flow of life? Is the true essence not so much a fixed object or attitude within the mind as a constant ongoing process, a creative movement toward integration that takes us through our lives?

Attempting to understand how we create ourselves, and the characters we read, resonates with our new understanding of the world, and its inherent complexity and ambiguity. Just as the electron cannot be captured within a single explanation, so too the self cannot be reduced to a single name; rather its essence lies in movement and integration.

The Science Story

Before we leave this issue of reading let us turn to another type of story—science itself. Science is that story our society tells itself about the cosmos. Science supposedly provides an objective account of the material world based upon measurement and quantification so that structure, process, movement, and transformation can be described mathematically in terms of fundamental laws.

Science proceeds by abstracting what is essential from the accidental details of matter and process. When Newton's apple falls it doesn't matter if it is ripe or green, a golden delicious or a Cox's orange pippin. Such qualities do not concern a science that prides itself on being value free. It does not matter if the person who measures the conductivity of copper, or the refractive index of quartz, is a Hindu, a born again Christian, or a staunch atheist. Neither does it matter if this experiment is carried out in a laboratory in Moscow, Delhi, or Chicago—the result will always be the same. Einstein's famous theory of relativity states that while phenomena appear different to someone close to a black hole, traveling close to the speed of light, or in a falling elevator here on earth, scientists in profoundly different environments will nevertheless always discover the same underlying laws of nature.

In this sense science appears to stand outside our earlier discussion of creative readings within a social and cultural context. Science asserts that the answers nature provides are independent of culture, belief, and personal values. Cultural relativism, it argues, has no place in science.

Certain aspects of this claim on the part of science may well be true but they miss an essential point. Science begins with our relationship to nature. The facts it discovers about the universe are answers to human questions and involve human-designed experiments. The Western scientific approach, for example, places nature in a series of highly artificial situations and demands that answers are given quantitatively—in terms of number.

Other societies, had they developed a strong science of matter and an associated technology, may have had quite a different relationship to the natural world. In turn, they would have asked other sorts of questions. They may have been more concerned with relationship, wholeness, the position of the human observer, and the role of consciousness in the world. They may have abstracted quantities or qualities different from those of, say, mass and velocity. This is not to say that a science created by Native Americans or Africans would in some way contradict or deny Western science. Rather it would provide a different framework for knowing the world. It would ask different questions and seek other sorts of answers. In this way alternative theories and types of explanations would be offered. In *Blackfoot Physics*[6] I attempted to portray such an alternative worldview and show that, while it is entirely consistent, it offers a different relationship to reality than that of Western science.

This is not to say, as some have erroneously argued, that one can choose to create any reality one wishes. Or that reality is no more than the expression of a particular belief system. Certainly objective aspects to the world clearly do exist, although different cultures may see these in different ways. No matter what you wish to believe you will still stub your toe if you kick a rock. No amount of cultural relativism will make a rock vanish or prevent a ripe apple from falling to the ground. On the other hand, the falling apple and the nature of the rock could play quite different roles in sciences of other cultural contexts.

Provided that such alternative approaches engage in disciplined argument and deduction, and that there is an element of careful attention to an observation, then the knowledge systems of other cultures have the right to stand as scientific viewpoints. It may be possible that other societies view the natural world through the prism of cooperation and symbiosis rather than environmental competition. Laws of nature may be seen as more organic than mechanical. Alternative sciences may be less concerned with prediction and control than with empathy and understanding.

Conceiving the possibility that alternative sciences could exist allows us to look back at Western science and ask how much of it is inevitable and objective, and how much is culturally conditioned and determined. To take one example, the desire for an ultimate level to

[6]Grand Rapids, Mich.: Phanes Press, 2002.

matter, and the need for a final solution that will provide closure to scientific questioning, does appear to be the manifestation of a persistent trait in Western civilization. This is not necessarily shared by other cultures that may be more willing to accept an infinity of qualitatively different levels and explanations that are forever open. Western novels are generally created around a logical scheme of development with a beginning, middle, and end. But many Arabic stories have no end and very little development. Classical Western music involves the alternation of tension and resolution and moves forward toward a coda in which everything is to be resolved and ended in a formal way. By contrast, Islamic music moves in a more inward way, not having any particular goal or ending, but rather opening into an infinity of variations between the various notes. The need for all-embracing explanations, fundamental levels, and definitive endings may therefore not be so much a characteristic of "science" itself but of a particular cultural mindset within the West.

In this sense science becomes the story that our civilization tells itself. It is a story about the universe, but told in such a way that it supports and gives credence to all that our society holds of value—analysis, prediction, technology, the accumulation of wealth and knowledge, the desire for control, progress, the need for closure and wrapping things up. Science adds credibility to our cultural dream by supporting it in a seemingly objective way.

Today we must be more willing to see Western science as a story that our society has been telling itself about the universe. To make such a statement is not to discredit science or to claim that "everything is relative." It is to argue that a story can be told in a variety of different ways. Science, as a story, is as great as any masterwork of art. It is a story that has produced enormous advances in several fields of knowledge. It is a story that has led to the triumphs of modern technology and has helped us to understand our world and ourselves. Nevertheless, certain aspects of this story have also led to the creation of problems that currently confront us.

When we say that science is a story told by the Western mind we must also remember that other cultures tell different stories. It is a new form of cultural imperialism to claim that the stories of other cultures are no more than myths that must be "corrected," exposed for their naiveté, or "made more scientific." Rather, they should be respected, for they represent different possible glances at the universe and different ways of structuring knowledge. If we take these various stories together they provide a rich multiplicity of perspectives, similar to those of a Cézanne painting.

The danger arises when a culture takes its own story as the absolute truth, and seeks to impose this truth on others as the yardstick for all knowledge and belief. We should never forget that, at their deepest level, all questions, all searches for knowledge, be they "scientific," "mystical," "philosophical," or "religious," point to a the same truth, but often in profoundly different ways.

For example, some years ago I had discussions with a Mohawk community about their school system. On the one hand, they knew the importance of youth remaining connected to their culture. On the other, they realized that if they were to survive in contact with the modern world, their community would have to confront "Western" science, medicine, and technology. To care for their land, for example, and deal with environmental threats, the next generation would need to know about biology, ecology, and chemistry.

So the community decided that mathematics and science should be team-taught with a teacher trained in Western science standing side by side with a Mohawk elder. When one spoke of molecules and chemical reactions, the other would discuss energies and relationships of powers and spirits. In this way children could learn the beauty and rationality of Western science but still remain in contact with the depth of their own tradition. Rather than Western science and indigenous cultures coming into conflict, each would enrich the other.

Where Do We Go from Here?

The natural world, and all those advances thought has created for us (from microeconomics to genetic engineering, from legal constitutions to satellites), are impelling us to take a step back, to pause and look around at all we have created. Our world is, in part, the actuality of all that is. But this actuality is perceived and shaped by thought. In turn, human thought, and the technologies it has produced, acts back on that world to transform it.

Human intervention within the natural world—the environment, our minds and bodies, means of travel and communication, creation of new materials, exploitation of resources, and so on—has had an enormous impact over the past 200 years. This impact continues at an accelerating rate. Now is the time to take stock. We need to look at our world, our different societies, and ourselves and ask where we are going. To do this during the first years of a new millennium seems particularly appropriate.

We must examine the structure of the various organizations we have created in the social, political, and religious fields. We must investigate our economic systems, the way various forms of governance operate, and the different ways people seek to help each other. We must ask if these structures and organizations continue to serve the purposes for which they were first created. Are they true to the spirit that once inspired them?

Earlier in this chapter we made a critical assessment of the Enlightenment dream. Yet the Enlightenment was also the period in which the best minds of a generation looked critically at society and governance. In England, the philosophers Hobbes and Locke enquired into the nature of human society and the maintenance of its good order. Voltaire and the Encyclopedists did something similar in France, while in the United States Jefferson and the founding fathers of the Constitution thought deeply about systems of governance. A similar surge of critical creativity is needed today.

Times have changed and new perspectives have emerged. Of key importance is our realization that we live on a finite planet, so that all our plans and actions resonate across the globe. The implications of any decision cannot be confined to one group, people, or nation. When the Iroquois needed to make a decision they thought of the implications for the seventh generation to come after them. Today we must think not only of the seventh generation of Americans, British, Germans, or Japanese but of peoples all over the world, rich or poor, industrialized or indigenous.

The need to pause is vital. The responsibility certainly falls upon the best minds of the planet, its philosophers, poets, artists, writers, politicians, and scientists. But it is also the responsibility of each one of us. For each of us intersects with this planet and its peoples in a wide variety of ways. If we are parents then we must think of our children, of the nature of their education, and of the type of work they will do. If we work, then we think of business, economics, the general future of work and leisure, and how we will generate fulfillment in our lives. When we go shopping, we place ourselves in contact with the entire planet, since food, clothing, and manufactured goods now come from every corner of the world. What are the implications of our choices, how does our new car or our evening meal affect people in Africa, Asia, or even small-town America?

When we step out into the street we are members of a community and we ask how communities form and flourish. We wonder about the deep need that is felt today to belong, to be part of some wider meaning, to be given the opportunity to contribute. And, when walking in nature, sitting quietly at home, or attending a place of worship we begin to think of the transcendent qualities of life and of all that is sacred.

In this book we have seen the various ways our thoughts can transform the world, and the ways consciousness changed during the twentieth century. Certainly the action of a pope or president can have great implications. But so too can our own individual thoughts. Each one of our thoughts changes the world in a tiny but subtle way. Multiply this by the tens of thousands of people in a town, the hundreds of millions in a country, the billions in the world, and human thought has an enormous impact.

Now we stand on the threshold of a pause. Each one of us is going to look out at the world and into his or her heart. Out of this creative suspension will come a new impulse. Each one of us will be responsible for that impulse, for that which is going to carry us forward into this millenium. The combination of our impulses, thoughts, and new attitudes will create a new world. To do so we will not only consider our own hearts but we will begin to dialogue with others, with nature, and with the sacred. We have left the dream of absolute certainty behind. In its place each of us must now take responsibility for the uncertain future.

POSTSCRIPT

This book was completed before the tragic events of September 11, 2001. Only in December of that year, as I was making my final corrections to the page proofs, did I reread the book's final sentence. It struck me as being both particularly ironic, in the light of what had happened, as well as being particularly true. Hence the need for this postscript.

Many felt that the world changed on September 11. In the immediate aftermath, things certainly appeared to have changed for the worse. We realized how fragile are so many of the social, economic, and international structures we have taken for granted. We noticed how much depends upon cooperation, good will, and collective sanity.

Another set of certainties seemed to have been stripped away. We could look to no external authority, no organization, no expert to guarantee our security and future prosperity. Again the answer, the responsibility, comes down to ourselves—individually, in families, community groups, and organizations from the local to the global levels. In this way we are called upon to examine the values and meanings of our lives. We ask, "What do we truly want for our world, and for the world of our children's children?" It is a question none of us can avoid.

Shortly after September 11, I made a trip to Spain and ended up in the town of Cordoba. Nine hundred years ago it was the most important city in the European world, far more significant than London or Paris. As the cultural center of Andalusia, it was the crown of Arabic culture, rivaling even Baghdad. It was while wandering through the streets of Cordoba that I learned of the great vision of harmony and learning that had once flourished in that part of the world, and still remains a dream for all of us.

Cordoba had its great Mosque, and philosophers of the caliber of Averroes and Ibn al-Arabi. It was the birthplace of the Jewish physician, scholar, and philosopher Moses Maimonides. Above all, Cordoba was a city devoted to learning and when power passed from the Arabs to the Catholic King Alfonso X, he continued their dream by creating a university where Jews, Christians, and Muslims could study side by side the arts of mathematics, astronomy, botany, philosophy, and medicine. Even their sacred books were translated into each other's languages.

Cordoba holds a lesson for us. It tells us that when people are united by a common respect for learning, by a love of the beauty and wonder of the world, and by a respect for the sacred spaces of others, then true brotherhood and sisterhood is possible. The deepest human values can transcend those divisions created by history and economics and reinforced by hatred and empty rhetoric.

I continue to believe in the future and, as I have argued in this book, I feel that we have been given an important chance. The crutches of certainty have been cast away as illusions. Together we, of whatever belief, history, or culture, must work together to create a common future —a future that respects the rights and aspirations of all people, values the spirit of learning, celebrates the values of beauty and truth, and cares for our planet's health.

Appendix

GÖDEL'S THEOREM

Earlier attempts to demonstrate the consistency or the completeness of mathematics used a system of symbolic logic to make statements about theorems and mathematical arguments. In other words, the status of mathematics was being examined and certified by a system that lay outside itself—by metamathematics. Gödel's approach was to develop a system that could make statements about itself. The system that talked about mathematics would itself be a part of mathematics, rather than lying outside the mathematics it discussed.

But how can a symbolic system be made to refer to itself? And how can metamathematics be incorporated into mathematics? Gödel's answer was ingenious in the extreme. His first step was to show that to every statement in metamathematics there corresponds a unique number. Since numbers are always part of mathematics then statements in metamathematics about theorems and their proofs can now be reduced to the manipulation of numbers in mathematics.

Step 1

Let us begin by arranging in a row all the symbols used in mathematics, along with all the symbols of logic and all the numbers. The list below is a great simplification but serves to explain the general idea of how things are going to work.

$$+ - \times = x\, y\, 0\, 1\, 2\, 3\, 4\, 5\, 6$$

Just as in the counting game place numbers below this row

+	−	×	=	x	y	0	1	2	3	4	5	6
1	2	3	4	5	6	7	8	9	10	11	12	13

If you give me the Gödel number 4, I now know that it stands for the equals sign, while the Gödel number 10 stands for the number 3.

Step 2

Not only can individual symbols and numbers be given a Gödel number, but an entire formula can be reduced to a number. Take as an example the formula 2 + 2 = 4. This involves the Gödel numbers 9 (which stands for the number 2), 1 (for the + sign), 4 (for equals), and 11 (for 4). We now add these numbers to get a new single Gödel number 34:

$$9 + 1 + 9 + 4 + 11 = 34$$

Thus the Gödel number 34 stands for the formula 2 + 2 = 4.

But now we run into a serious problem, for 34 is also the number for the formula 3 + 1 = 4. At this stage what we have been taking for Gödel numbers are not unique and, when we deal with more complicated formulas, one Gödel number can stand for a number of different formulas. Clearly this will not suffice to define mathematical formulas in a unique way.

It was at this point that Gödel proposed using prime numbers. (A prime number, such as 7, 11, 13, 23, etc., cannot be factored into other numbers, whereas a nonprime such as 12 can be factored into 2×6, and 3×4.)

Thus the new Gödel number for $2 + 2 = 4$ now becomes (using the sequence 9, 1, 9, 4, 11):

$$2^9 \times 3^1 \times 5^9 \times 11^4 \times 13^{11}$$

This (very large) number is unique and is shared by no other formula. Thus could Gödel express every symbol and every formula in mathematics by a unique Gödel number.

Step 3

What applies to a single formula also applies to a theorem and its proof. Simply assign numbers to each of the symbols and to each line in the proof. Now work out the Gödel number for this proof *Y*. We can now say that the theorem with Gödel number *X* has a proof with Gödel number *Y*. This is written down as Dem(*Y*, *X*).

But it may also be the case that some sequence *Y* does not prove the truth of the theorem *X*. Gödel wrote this as ~Dem(*Y*,*X*). Clearly if both Dem(*Y*,*X*) and ~Dem(*Y*,*X*) could be shown to coexist then mathematics would be contradictory.

Step 4

We have come quite a distance in the argument for we can now reduce all the theorems of mathematics and all their proofs to a series of Gödel numbers. Yet we still have not shown how a logical system can actually refer to itself. One further turn of the logical screw is necessary.

Suppose we take one of the basic facts of mathematics—that the numbers go on forever so that every number *y* has a successor *x*. Mathematicians like to phrase this in the following way, "There exists some number *x*, such that *x* is the successor of *y*." Written as a formula this reads

$$(\exists x)(x = sy) \qquad (1)$$

The number y has, as we have seen above, the Gödel number 6. In addition, the entire formula has its own Gödel number that can be calculated like any other Gödel number. Call the number of the formula above M.

In turn we can put this number M back into the formula in place of the number 6 (that is, the variable y).

$$(\exists x)(x = sM) \qquad (2)$$

This formula says "there exists some number x such that it is the successor to the number M." Admittedly M is a Gödel number but as a number in arithmetic it is no different from the number 6 (which stands for y).

As with every other formula in mathematics, line (2) has a Gödel number that can be calculated. But there is a second way to calculate that number. We can calculate it in such a way that metamathematics begin to mirror and reflect each other.

Suppose we write down the statement: "The formula obtained from the formula whose number is 'M' when you substitute the number M for the variable with number 6."—Statement (a)

Statement (a) is unambiguous. It's a statement that can be written down in symbolic form so that its Gödel number, N, can be calculated. In other words the statement with Gödel number N (Statement a) and line (2) are mirror images of each other—mathematics and metamathematics now reflect each other within the same system.

Having achieved this result—mirroring metamathematical statements within mathematics itself—Gödel could go on to construct the statement: "The formula with Gödel number Z is not demonstrable."—Statement (b)

In other words Gödel had constructed a statement of the type "I am not demonstrable" or "I cannot be proved."

Gödel also added two final steps to the argument. First, he showed that although Statement (b) cannot be demonstrated within his system, it is nevertheless a true statement. In other words he had shown the truth of the statement, "Mathematics is incomplete."

Second, he took a step that is reminiscent of the coexisting statements Dem(*Y*, *X*) and ~Dem(*Y*, *X*) that would imply that mathematics is not consistent.

What he showed was that, if Statement (b) were in fact demonstrable, it would also follow that the negation of statement (b) would be also demonstrable. That is, both a statement and its negation would simultaneously be demonstrable. But if statement (b) would be demonstrable this would mean that mathematics is complete—that is, every statement can be demonstrated. Hence Gödel's second conclusion, "If mathematics were *complete*—that is, if statement (b) could be demonstrated—then it would be *inconsistent*."

C

D

E

F

G

H

I

J

K

L

M

S

T

U